Macromolecular Symposia 214

6th Annual UNESCO School & IUPAC Conference on Macromolecules & Materials Science

Berg-en-Dal, South Africa
April 14–17, 2003

Symposium Editors:
R. D. Sanderson, H. Pasch, Stellenbosch, South Africa

pp. 1–379 · August 2004
ISBN 3-527-31047-9

Macromolecular Symposia publishes lectures given at international symposia and is issued irregularly, with normally 14 volumes published per year. For each symposium volume, an Editor is appointed. The articles are peer-reviewed. The journal is produced by photo-offset lithography directly from the authors' typescripts.
Further information for authors can be found at http://www.ms-journal.de
Suggestions or proposals for conferences or symposia to be covered in this series should also be sent to the Editorial office (E-mail: macro-symp@wiley-vch.de).

Macromolecular Symposia:
Annual subscription rates 2005
Macromolecular Full Package: including Macromolecular Chemistry & Physics (24 issues), Macromolecular Rapid Communications (24), Macromolecular Bioscience (12), Macromolecular Theory & Simulations (9), Macromolecular Materials and Engineering (12), Macromolecular Symposia (14):

Europe	Euro	7.088 / 7.797
Switzerland	Sfr	12.448 / 13.693
All other areas	US$	8.898 / 9.788

print only **or** electronic only / print **and** electronic

Postage and handling charges included. All Wiley-VCH prices are exclusive of VAT. Prices are subject to change.

Single issues and back copies are available. Please ask for details at: service@wiley-vch.de

Orders may be placed through your bookseller or directly at the publishers:
WILEY-VCH Verlag GmbH & Co. KGaA, P. O. Box 10 11 61, 69451 Weinheim, Germany, Tel. +49 (0) 62 01/6 06-400, Fax +49 (0) 62 01/60 61 84, E-mail: service@wiley-vch.de

---

For USA and Canada: Macromolecular Symposia (ISSN 1022-1360) is published with 14 volumes per year by WILEY-VCH Verlag GmbH & Co. KGaA, Boschstr. 12, 69451 Weinheim, Germany. Air freight and mailing in the USA by Publications Expediting Inc., 200 Meacham Ave., Elmont, NY 11003, USA. Application to mail at Periodicals Postage rate is pending at Jamaica, NY 11431, USA. POSTMASTER please send address changes to: Macromolecular Symposia, c/o Wiley-VCH, III River Street, Hoboken, NJ 07030, USA.

Macromolecular Symposia

Articles published on the web will appear several weeks before the print edition. They are available through:

www.ms-journal.de

www.interscience.wiley.com

6th Annual UNESCO School & IUPAC Conference on Macromolecules & Materials Science Berg-en-Dal (South Africa), 2003

Preface
H. Pasch

1. Persistence Length of Cartilage Aggrecan Macromolecules Measured via Atomic Force Microscopy. 1
   Laurel Ng, Alan J. Grodzinsky, John Sandy, Anna Plaas, Christine Ortiz*

2. Polymer Morphology: A Guide to Macromolecular Self-Organization . . . 5
   D. C. Bassett

3. Crazing and Fracture in Polymers: Micro-Mechanisms and Effect of Molecular Variables 17
   H. H. Kausch,* J.-L. Halary, C. J. G. Plummer

4. Environmental Stress Cracking of Polymers Monitored by Fatigue Crack Growth Experiments 31
   Volker Altstaedt,* Sven Keiter, Michael Renner, Alois Schlarb

5. Toughness Enhancement of Nanostructured Amorphous and Semicrystalline Polymers 47
   Georg H. Michler,* Rameshwar Adhikari, Sven Henning

6. Nanostructures and Functionalities in Polymer Thin Films. 73
   Manfred Stamm,* Sergiy Minko,* Igor Tokarev, Amir Fahmi, Denys Usov

7. Structure-Property Relationships in Rubber-Modified Styrenic Polymers . 85
   Walter Heckmann,* Graham Edmund McKee, Falko Ramsteiner

8. Microdeformation in Heterogeneous Polymers, Revealed by Electron Microscopy 97
   Christopher J. G. Plummer,* Philippe Béguelin, Chrystelle Grein, Rudolph Gensler, Laure Dupuits, Cedric Gaillard, Pierre Stadelmann, Hans-Henning Kausch, Jan-Anders E. Månson

9. Morphology and Properties of Particulate Filled Polymers. 115
   Béla Pukánszky,* János Móczó

10. Morphology and Properties of Poly(oxymethylene)
    Engineering Plastics 135
    Tatiana Sukhanova,* Vladimir Bershtein, Mimi Keating,
    Galina Matveeva, Milana Vylegzhanina, Victor Egorov,
    Nina Peschanskaya, Pavel Yakushev, Edmund Flexman, Stefan Greulich,
    Bryan Sauer, Kathleen Schodt

11. Linear Viscoelasticity and Non-Linear Elasticity of Block Copolymer
    Blends Used as Soft Adhesives. 147
    Alexandra Roos, Costantino Creton*

12. Micromechanical Mechanisms for Toughness Enhancement in
    β-Modified Polypropylene. 157
    Sven Henning, Rameshwar Adhikari, Goerg H. Michler,*
    Francisco J. Baltá Calleja, József Karger-Kocsis

13. Morphology and Micromechanical Behaviour of SBS Block Copolymer
    Systems 173
    R. Adhikari, R. Godehardt, W. Lebek, S. Goerlitz, G. H. Michler,*
    K. Knoll

14. Conductive Composite Materials of Polyethylene and Polypyrrole with
    High Modulus and High Strength 197
    Dan Zhu, Yuezhen Bin, Kumiko Oishi, Yasuyuki Fukuda,
    Takahiro Nakaoki, Masaru Matsuo*

15. Use of the Surfmer 11-(Methacryloyloxy)undecanylsulfate MET as a
    Comonomer in Polystyrene and Poly(methyl methacrylate) 217
    P. C. Hartmann, A. Pienaar,* H. Pasch,* R. D. Sanderson

16. Strain-Controlled Tensile Deformation Behavior and Relaxation Proper-
    ties of Isotactic Poly(1-butene) and Its Ethylene Copolymers 231
    Mahmoud Al-Hussein, Gert Strobl*

17. Technology and Stress Relaxation of Biaxially Oriented Polyolefin Shrink
    Films 241
    Arthur Bobovitch,* Yakov Unigovski, Albert Jarashneli,
    Emmanuel M. Gutman

18. Failure Envelope Curves in Polyethylene Solids. 251
    Koh-hei Nitta,* Takashi Ishiburo

19. Thermal Oxidation and Its Relation to Chemiluminescence from
    Polyolefins and Polyamides. 261
    Lyda Matisová-Rychlá,* Jozef Rychlý, P. Tiemblo,
    J. M. Gómez-Elvira, M. Elvira

20. Influence of Reactive Compatibilization on the Morphology of
 Polypropylene/Polystyrene Blends 279
 J. Pionteck,* P. Pötschke, U. Schulze, N. Proske, A. Kaya,
 H. Zhao, H. Malz

21. Side Chain Extension of Polypropylene by Aliphatic Diamine
 and Isocyanate. 289
 K. Y. Kim, S. C. Kim*

22. New Approaches for the Development of Highly Stable
 Polypropylene 299
 Minoru Terano,* Boping Liu, Hisayuki Nakatani

23. Structure, Dynamics and Properties of Materials with Polymers
 Having Complex Architectures 307
 Tadeusz Pakula

24. Morphology and Temperature Phase Transitions in α,ω-Alkanediols
 with Different Chain Lengths 317
 Vyacheslav Marikhin,* Victor Egorov, Elena Ivan'kova,
 Liubov Myasnikova, Elena Radovanova, Boris Volchek,
 Darya Medvedeva, Alan Jonas

25. Synthesis, Characterization and Properties of Novel
 Poly(Ester-Amide-Urethane)s 339
 Shahram Mehdipour-Ataei,* Parvin Einollahy

26. Self-Diffusion of PEO-Modified Paclitaxel in Aqueous Solution:
 Hydrodynamic Properties 351
 Michael Hess,* Manfred Zähres, Byung-Wook Jo

27. The Use of Pressure-Volume-Temperature Measurements in Polymer
 Science 361
 Michael Hess

Author Index

Adhikari, Rameshwar . 47, 157, 173
Al-Hussein, Mahmoud 231
Altstaedt, Volker 31
Baltá Calleja, Francisco J. . . 157
Bassett, D. C. 5
Béguelin, Philippe 97
Bershtein, Vladimir 135
Bin, Yuezhen 197
Bobovitch, Arthur 241
Creton, Costantino 147
Dupuits, Laure 97
Egorov, Victor 135, 317
Einollahy, Parvin 339
Elvira, M. 261
Fahmi, Amir 73
Flexman, Edmund 135
Fukuda, Yasuyuki 197
Gaillard, Cedric 97
Gensler, Rudolph 97
Godehardt, R. 173
Goerlitz, S. 173
Gómez-Elvira, J. M. 261
Grein, Chrystelle 97
Greulich, Stefan 135
Grodzinsky, Alan J. 1
Gutman, Emmanuel M. 241
Halary, J.-L. 17
Hartmann, P. C. 217
Heckmann, Walter 85
Henning, Sven 47, 157
Hess, Michael 351, 361
Ishiburo, Takashi 251
Ivan'kova, Elena 317
Jarashneli, Albert 241
Jo, Byung-Wook 351
Jonas, Alan 317
Karger-Kocsis, József 157
Kausch, Hans-Henning . . 17, 97
Kaya, A. 279
Keating, Mimi 135
Keiter, Sven 31
Kim, K. Y. 289

Kim, S. C. 289
Knoll, K. 173
Lebek, W. 173
Liu, Boping 299
Malz, H. 279
Månson, Jan-Anders E. 97
Marikhin, Vyacheslav 317
Matisová-Rychlá, Lyda 261
Matsuo, Masaru 197
Matveeva, Galina 135
McKee, Graham Edmund . . . 85
Medvedeva, Darya 317
Mehdipour-Ataei, Shahram . . 339
Michler, Goerg H. . . . 47, 157, 173
Minko, Sergiy 73
Móczó, János 115
Myasnikova, Liubov 317
Nakaoki, Takahiro 197
Nakatani, Hisayuki 299
Ng, Laurel 1
Nitta, Koh-hei 251
Oishi, Kumiko 197
Ortiz, Christine 1
Pakula, Tadeusz 307
Pasch, H. 217
Peschanskaya, Nina 135
Pienaar, A. 217
Pionteck, J. 279
Plaas, Anna 1
Plummer, Christopher J. G. . . 17, 97
Pötschke, P. 279
Proske, N. 279
Pukánszky, Béla 115
Radovanova, Elena 317
Ramsteiner, Falko 85
Renner, Michael 31
Roos, Alexandra 147
Rychlý, Jozef 261
Sanderson, R. D. 217
Sandy, John 1
Sauer, Bryan 135
Schlarb, Alois 31

Schodt, Kathleen   135
Schulze, U.   279
Stadelmann, Pierre   97
Stamm, Manfred   73
Strobl, Gert   231
Sukhanova, Tatiana   135
Terano, Minoru   299
Tiemblo, P.   261
Tokarev, Igor   73

Unigovski, Yakov   241
Usov, Denys   73
Volchek, Boris   317
Vylegzhanina, Milana   135
Yakushev, Pavel   135
Zähres, Manfred   351
Zhao, H.   279
Zhu, Dan   197

Preface

UNESCO Chemistry for Life Division in Paris has awarded 13 Associated Centres for research in chemical science and education. One of these is the UNESCO Associated Centre for Macromolecules & Materials which is part of the Chemistry Department of the University of Stellenbosch, South Africa.

As part of its activities, UNESCO encourages and sponsors UNESCO Schools and scientific conferences in collaboration with scientifc associations such as the International Union for Pure and Applied Chemistry (IUPAC). The UNESCO School and Conference on Macromolecules & Materials Science is held annually at different locations in South Africa (website: http://www.sun.ac.za/unesco/unesco.htm). World authorities in various fields of macromolecular science are invited to give tutorials at the UNESCO School and informative plenaries at the conference. The exposure to new ideas and advanced concepts in macromolecular science is of great importance for South African students and senior staff alike coming from different universities and research institutions. It is particularly valuable that with the support of UNESCO generous concessions can be made for attendees from disadvantaged communities and from countries with emerging technologies. The 6th UNESCO School & IUPAC Conference focused on polymer properties with a special session on the characterization of polyolefins. Abridged versions of a number of papers are compiled to create the present volume of Macromolecular Symposia. The content of the papers is also available in the Virtual Teaching Encyclopaedia which contains papers from previous UNESCO conferences as well (at the above website).

H. Pasch

Macromol. Symp. **2004**, *214*, 1-4

Persistence Length of Cartilage Aggrecan Macromolecules Measured via Atomic Force Microscopy

Laurel Ng,[a] *Alan J. Grodzinsky,*[a,b,c] *John Sandy,*[e] *Anna Plaas,*[f]

*Christine Ortiz**[d]

[a] Biological Engineering Division, Massachusetts Institute of Technology, 77 Massachusetts Avenue, Cambridge, MA 02114 USA
[b] Departments of Electrical Engineering and Computer Science, Massachusetts Institute of Technology, 77 Massachusetts Avenue, Cambridge, MA 02114 USA
[c] Mechanical Engineering, Massachusetts Institute of Technology, 77 Massachusetts Avenue, Cambridge, MA 02114 USA
[d] Materials Science and Engineering, Massachusetts Institute of Technology, 77 Massachusetts Avenue, Cambridge, MA 02114 USA
E-mail : cortiz@mit.edu
[e] Shriner's Hospital, Tampa, and [f] Departments of Pharmacology and Therapeutics and Internal Medicine, University of South Florida, Tampa, FL USA 33620

Summary: Tapping mode atomic force microscopy (TMAFM) was employed to directly calculate the persistence length of individual fetal bovine epiphyseal and mature nasal cartilage aggrecan monomers, as well as their constituent chondroitin sulfate glycosaminoglycan chains.

Keywords: aggrecan; atomic force microscopy; biopolymers; cartilage; glycosaminoglycan

Introduction

Cartilage is a highly specialized, dense connective tissue found between the surfaces of movable articular joints whose main function is to bear stresses during joint motion. The negatively charged disaccharide *chondroitin sulfate glycosaminoglycan* (CS-GAG) macromolecules are a major determinant of the tissue's ability to resist compressive and shear loading *in vivo*, e.g., responsible for >50% of the equilibrium compressive elastic modulus under normal physiological conditions (0.15 M salt concentration).[1] Approximately 100 CS-GAGs are covalently bound at extremely high densities (separated by ~2-4 nm) to a 250 kDa core protein forming the *aggrecan* molecule. Multiple aggrecan molecules self-assemble to form supramolecular *proteoglycan*

© 2004 International Union of Pure and Applied Chemistry   DOI: 10.1002/masy.200451001

aggregates by noncovalently end-attaching to a *hyaluronan* (HA) central filament, an interaction that is stabilized by the binding with *link protein*. These aggregates form a gel-like component enmeshed within a network of reinforcing collagen fibrils.

Experimental

Purified A1A1D1D1 aggrecan monomers from the fetal bovine epiphyseal growth plate region and mature bovine nasal region was dialyzed first against 500 volumes of 1 M NaCl and then against HOH to remove excess salts to 5-25 μg/mL. A positively-charged amine-functionalized muscovite mica surface (Pelco International, Redding, CA) was prepared with a treatment of 60 μL 0.01% 3-amino-propyltriethoxysilane (APS) for 20-30 min (Sigma Aldrich Co., St. Louis, MO) v/v MilliQ water (Millipore Corp, Bedford, MA.). 60 μL of 250 μL/mL aggrecan solution was allowed to incubate on the APS-mica for 20-30 min, then rinsed, and dried in air for 1 h before imaging in tapping mode in ambient conditions with a Nanoscope IIIa Multimode atomic force microscope (TMAFM) (Digital Instruments (DI), Santa Barbara, CA) and Olympus AC240TS-2 Si cantilevers (probe tip radius < 10 nm, spring constant 2 N/m).

Results and Discussion

TMAFM visualization of partially hydrated individual aggrecan monomers with unprecedented clarity and molecular-level resolution (Figure 1a) enables the direct quantitative measurement of the ultrastructure, dimensions, and conformation of both the protein core backbone and individual CS-GAG chains.

To calculate an effective persistence length, L_p, a series of equal length vectors were iteratively projected onto the digitized trace of the core protein and GAG contours in increments of l =1.2 nm (Figure 1b). For each l, the bend angle with respect to the previous vector, θ, was calculated for each vector (i.e. as a function of position along the polymer chain) yielding $<\theta^2(l)>$. Using the assumptions of the Worm-Like Chain (WLC) model, L_p was estimated from the linear relationship of the variance of θ as a function of l (Figure 2a). The kurtosis of θ at each l was plotted to assess the WLC behavior of the molecule (Figure 2b). Table 1 summarizes the results for L_p, the chain end-to-end distance (R_{ee}), the trace length (L_c), the aggrecan diameter (w), and GAG-GAG interchain spacing (d).

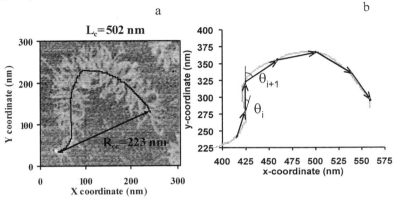

Figure 1. a) TMAFM height image taken of an aggrecan molecule and b) projection of a series of equal length vectors onto the digitized trace of the core protein backbone of the aggrecan molecule shown in a)

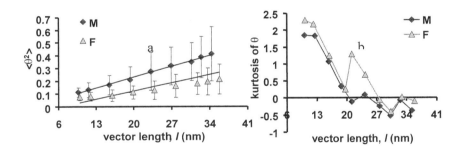

Figure 2. a) Calculation of L_p and b) kurtosis of θ both for aggrecan $n_{mature}=15$ (M), $n_{fetal}=15$ (F)

Table 1.   Properties of aggrecan protein core and CS-GAG chains measured by TMAFM

cartilage	aggrecan protein core				CS-GAG			
units=nm	L_c	R_{ee}	L_p	w	L_c	R_{ee}	L_p	d
mature nasal	353±88	226±81	82	47±12	32±5	26±7	14	4.4±1.2
	n=40	n=40	n=15	n=108	n=49	n=49	n=17	n=40
fetal epiphyseal	398±57	257±87	110	57±14	41±7	32±8	21	3.2±0.8
	n=102	n=102	n=15	n=104	n=102	n=102	n=21	n=102

n = number of molecules used in calculation

Conclusion

Distinct differences in the nanoscale properties between two aggrecan populations (mature nasal versus fetal epiphyseal) have been clearly observed via TMAFM imaging. Hence, it is clear that such studies, as a function of age, disease, and injury, have great potential to yield new insights into the molecular origins of cartilage dysfunction.

[1]   M.D. Buschmann, A.J. Grodzinsky, *J. Biomech. Eng.* **1995**, *117*, 179.

Macromol. Symp. **2004**, *214*, 5-15

Polymer Morphology: A Guide to Macromolecular Self-Organization

D. C. Bassett

JJ Thomson Physical Laboratory, University of Reading, Reading RG6 6AF UK

E-mail: d.c.bassett@rdg.ac.uk

Summary: The study of polymer morphology continues to be the principal means of acquiring knowledge and understanding of macromolecular self-organization. Long-standing problems of the nature of melt-crystallized lamellae and spherulitic growth have been resolved, bringing understanding of how characteristic properties such as a broad melting range and spatially-varying mechanical response are inherent in spherulitic morphologies. This reflects the distinctive features of the crystallization of long molecules, i.e. that they impede each other and, for faster growth, form rough basal surfaces. Knowledge of morphology is an essential accompaniment to the informed development of advanced polymeric materials and a full understanding of their structure/property relations.

Keywords: crystallization; fibers; macromolecular self-organization; polymer morphology; structure/property relationships

Introduction

Crystalline polymers are important and successful materials with ever-widening uses, especially polyethylene and polypropylene which are the two in greatest production. Their usefulness is derived not only from their characteristic polymeric properties of low modulus elasticity, easy processability and toughness, but also because these can be substantially enhanced by suitable treatments. This goes beyond simple orientation of the long molecules, as in fibres, and concerns modifications of the internal microstructure. In crystalline polymers this is lamellar with long molecules folded, to a degree, back and forth between the fold surfaces. Contrary to early opinion, properties are not primarily dependent on the degree of crystallinity but upon the way lamellae are organized within a material which, for polymers crystallized from a quiescent melt, is characteristically spherulitic.

DOI: 10.1002/masy.200451002

The study of this internal organization, primarily using microscopes – with diffraction techniques used in a complementary role to quantify a defined textural model – is the subject of polymer morphology. This has been and continues to be the principal means of establishing knowledge and understanding of how long molecules self-organize, essentially because the microstructure contains a uniquely detailed record of the history of a polymeric sample, notably its crystallization and deformation, as well as indications of its molecular characteristics. Polymer morphology is key to understanding the fundamental aspects of macromolecular self-organization across the length scales from molecular through lamellar to spherulitic dimensions, and essential for informed control of processing to optimize properties. Following recent advances, the formation of spherulites and melt-crystallized lamellae are now much better understood, although the nature of the molecular networks underlying these entities remains to be well characterized, lying beyond currently achievable resolutions. The value, role and achievements of polymer morphology are the concern of this paper which is complementary to recent reviews.[1-3]

The Morphological Hierarchy

Examination of the interior of a crystalline polymer (of which polyethylene is the archetype, being both the simplest flexible macromolecule and economically the most important) reveals texture at a range of dimensional levels, depending on the magnification and resolution of the imaging technique used. Optical microscopy, the only available means until the advent of modern electron microscopes in the mid 1950s, shows inhomogeneity, at the μm level and above, in the form of spherulites. Literally little spheres, these are polycrystalline with equivalent radii because they contain chainfolded lamellae growing radially outwards.

The discovery of lamellae and their chainfolded constitution in 1957 marked the beginning of modern understanding of macromolecular self-organization. The, then newly synthesized, Ziegler-Natta linear polyethylene forms thin lamellae ~10 nm thick when precipitated from dilute solution in xylene. Electron diffraction showed that the ~1 μm long molecules were transverse within lamellae, whence it was inferred that they must be repetitively folded, back and forth at the basal or fold surfaces.[4] Chainfolded lamellae are a general feature of crystalline polymers, crystallized both from solution and melt, occurring because they are the fastest way in which long chains can crystallize. Their thickness is a physically, rather than chemically, determined quantity, increasing with crystallization temperature. This is to be expected because the gain in volume free energy on crystallization is linear in inverse supercooling, decreasing to zero at the equilibrium melting point.

The presence of fold surfaces increases the free energy of a lamella, making it thermodynamically unstable with respect to an increase of thickness as this decreases the proportion of fold surfaces in the system. Thickening can be accomplished in two ways. The first is by so-called *partial melting* in which the thinnest lamellae in a sample with a range of thicknesses – as would result from non-isothermal crystallization on cooling – melt, then recrystallize on their solid surroundings at an increased thickness due to the higher crystallization temperature. This occurs for solution-crystallized lamellae at temperatures above their crystallization temperature but melt-crystallized lamellae typically show *isothermal lamellar thickening* at their (somewhat higher) crystallization temperatures. In this case, thickening occurs via the strong characteristic longitudinal vibrations. Recent research, discussed below, has shown that thickening of a single polyethylene lamella occurs both at the growing edge, whose thickness increases with radial distance, and in the interior; the latter implies that new material must be incorporated into the lamella behind the growth interface.

The ubiquitous presence of lamellae in crystalline polymers lies behind characteristic properties, first, a reduced and broad melting range, the breadth linked to the spread of lamellar thickness and second, their low modulus. Fold surfaces interrupt the continuous sequence of covalent bonds along the **c** axis of polymeric crystal structures, introducing weaker van der Waals' bonding between lamellae[5, 6]. The actual modulus of a crystalline polymer specimen depends upon the details of the relative arrangement of covalent and van der Waals' bonds. It is least when they are in series and greatest when they are in parallel, with each weak interface both narrow and bridged laterally by a crystallite. Based on these considerations, manipulation of the morphology by suitable processing has produced a number of commercial high-modulus polyethylene fibres with moduli moving towards the theoretical limit ~ 3 Mbar, some 50% above that of steel.

The mutual arrangement and interconnectedness of lamellae is crucial in determining properties. In specimens crystallized from a quiescent melt the texture is characteristically spherulitic. Once it became possible to examine the internal lamellar organization, using chlorosulphonation of polyethylene and permanganic etching of polyolefines and other polymers, it was found that homopolymer spherulites have a dominant/subsidiary microstructure. Individual first-forming or dominant lamellae branch periodically then splay apart, establishing a framework which is filled in by later-forming subsidiary lamellae. Within spherulites, therefore, there is an intrinsic variation in lamellar texture and properties on the scale of the inter-dominant spacing.

This variation may differ considerably. There are differences in orientation between dominant and subsidiary lamellae, giving them different responses to applied stress[7] for an isothermally-crystallized homopolymer of low polydispersity, possibly coupled with small differences in their thickness due to differential isothermal thickening. In more polydisperse homopolymers and, especially, random copolymers, fractional crystallization may place different molecular species in different locations. In the former materials, shorter polyethylene or poorly tactic i-polypropylene molecules, which are prone to crystallize later, may become concentrated between dominant lamellae and/or in inter-spherulite boundaries, promoting brittleness due to concentrations of shorter molecules. Similarly, in random polyethylene copolymers, more-branched molecules (correlated with shorter lengths for Ziegler-Natta synthesis) may concentrate in later-crystallizing regions with comparable consequences.

Copolymers bring additional complexity. Greater concentrations of comonomer lower the equilibrium melting point, reducing the isothermal supercooling and thence the growth rate, an effect which, if sufficiently large, can render growth morphologically unstable with additional microstructural consequences. If, in addition, crystallization is not isothermal, this will increase the supercooling and reduce the thickness of later-crystallizing lamellae, giving them a lower melting point. Spherulitic textures, therefore, go far beyond the evident differences in radial orientation to include systematic variations of lamellar orientation, thickness and molecular composition on the scale of the inter-dominant separation.

Such matters are further complicated when crystallization occurs under applied stress or flow as in industrial processing. Principal of these is the change from effectively point nucleation of spherulites to extended linear nuclei forming the backbone of so-called shish-kebab or, preferably, row structures. Although the growth processes at the crystal/melt interface are presumably the same, linear nucleation does bring significant changes. First, the backbone or linear nucleus, which forms to sustain the applied stress, is higher melting than the thin lamellae which grow from it; in i-polypropylene the difference may be ~50 K. Second, the different geometry means that lamellae encounter different growth conditions to those in spherulites and respond accordingly. These are principally two different conditions. One is that lamellae growing in a close-packed parallel array are unable to thicken uniformly, unlike those which diverge from their nearest neighbours. In practice, those lamellae in such an array which thicken do so at the expense of others which fail to propagate.[8] A striking consequence is that whereas the thickest and highest melting lamellae in spherulites are the oldest, at the centre, polyethylene shish-kebabs melt first not only at the outside of a row but also adjacent to the extended nucleus.[9] The second consequence is that diffusion conditions differ at the growth interface for molecular species segregated during growth. The greater space between lamellae which diverge, compared to those which grow parallel, reduces the local concentration of segregants. If these lower the equilibrium melting point, as in polyethylene random copolymers, the local supercooling will decrease, giving row structures a slower radial growth rate than adjacent spherulites.[9] In turn, if the growth-rate depression becomes sufficiently large, it leads to constitutional supercooling, and the growth interface becomes unstable towards faster-growing protuberances. In row structures this becomes evident by the trace of the interface with the melt, in a section through and parallel to the nucleus, changing from linear to quasi-sinusoidal.[10] Such changes are sensitive to the nature of the segregated species. There are differences, for example, between polyethylene copolymers of the same branching ratio synthesized by Ziegler-Natta means as opposed to the use of single-site or metallocene catalysts.[11]

The spatial variations inherent in spherulitic textures may affect properties in undesirable ways. For example, it is difficult to prepare clear films of i-polypropylene because of light scattering from large spherulites whose size is controlled by the number of primary nuclei. While this is easily increased for polyethylene, achieving clarity in i-polypropylene remains a persistent problem. Internal textures are affected principally by primary nucleation density, molecular constitution and crystallization temperature, variables which offer a limited degree of control.

Spherulitic Growth

The key feature which causes growth to be spherulitic rather than single-crystalline in polymers is that lamellae periodically branch, then diverge. Iteration of this procedure gives spherulites with all radii having the same growth axis, **b** for polyethylene, **a*** for i-polypropylene. Their straight traces show that adjacent lamellae diverge because they are repelled by a short-the next branch. Branching is often, but not necessarily, at giant screw dislocations whose Burgers' vector is the lamellar thickness measured along **c**, the chain axis. It has now been shown in considerable detail for model systems of monodisperse long n-alkanes that the repulsion is due to molecular cilia, both transient and permanent, i.e. uncrystallized portions of molecules partly attached to a lamella. The divergence of adjacent lamellae increases for longer cilia and thinner lamellae.[12] Transient cilia must be present whenever a long molecule adds to a lamella in stages; permanent cilia exist in co-crystallizing binary blends. The effects of transient and permanent ciliation have been separated, with the latter disappearing in blends when the guest molecule is twice as long as the host and crystallizes in a hairpin conformation in a lamella of uniform thickness.[13,14] These studies also revealed a second cause of lamellar divergence, namely surface roughness - a feature associated with rapid crystallization.

Surface Organization and Reorganization

The formation of initially rough surfaces, prone to reorganization, has recently been shown to be characteristic of rapid polymeric crystallization from observations down **b**, the growth direction and radius of polyethylene row structures developing from linear nuclei. These relate to the lamellar habits adopted in different circumstances. In melt-crystallized linear polyethylene, fold surfaces are 201 at low supercoolings, inclined to the chain axis **c** by ~35° but, at ≤~126 °C for typical molecular weight, dominant lamellae viewed down **b,** have S- or C-profiles inclined at $\leq\sim35^{\circ}$ to **c** in their centres. Dominant lamellae of the second kind are found in banded spherulites; banding and twisted growth does not occur for the first kind of lamellae.

The different profiles come about because at lower temperatures the next molecular layer is added to a lamella before folds in the previous one have had time to attain the preferred 201 packing. Conversely at higher temperatures, with slower growth, there is time for this optimum fold packing to be adopted before the next layer is added.[15] The reason for *inclined* 201, being adopted rather than *perpendicular* 001 fold surfaces (with respect to the chain axis) is that they provide more surface area per fold. Inclined basal surfaces are found not only in polyethylene but also in the n-alkanes, which have the same subcell whose low cross-sectional area per chain militates against too close a packing of bulky end groups or folds. However, the reduction in free energy from fold surface ordering must be small in relation to that from the crystallization of fold stems. Moreover, the latter increases with supercooling whereas the energetic cost of a (disordered) fold surface will be more nearly constant and independent of temperature. It is to be expected, therefore, that faster crystallization will initially give disordered surfaces while ordered surfaces will only occur for slower growth; this is the case even for monodisperse long n-alkanes.

Study of the growth of polyethylene row structures from highly oriented fibres as linear nuclei, has not only given substance to these expectations but also revealed new aspects of macromolecular self-organization. For slower growth, $\geq 127\ ^{\circ}$C for typical linear polymers, lamellae do indeed from inclined on the linear nucleus, the double 201 orientation presenting a chevron appearance viewed down the **b** axis. But, beyond a certain radial distance this gives way to all lamellae lying normal to the nucleating fibre, yet retaining their 201 surfaces. This new phenomenon has been ascribed to the effect of internuclear interference when one molecule is able to crystallize on adjacent lamellae simultaneously.[16] This effect will disappear and allow faster radial growth when lamellae are parallel and sufficiently separated.

By contrast, for faster growth, lamellae form perpendicular to the nucleus, with 001 surfaces. Then, with increasing radial distance, they separate (with some failing to propagate), thicken, and develop S-and C-profiles together with banded growth. The latter two are a consequence of delayed fold-surface ordering within the constraints of an existing lamella to reduce the surface stress of disordered fold packing. The adoption of an inclined profile in the central, oldest region of a lamella will, to maintain integrity, give an S-profile, or a C-profile in narrower, oblique cross-sections.

Confirmation that it is the (partial) relief of fold surface stress which drives banded growth in polymers comes from the development of row structures in random copolymers of polyethylene. Slower growth at higher temperatures begins exactly as for rows of the linear polymer, but in a range now lowered because of the inherently slower growth. Lamellae are inclined, with an initial chevron pattern which gives way to inclined lamellae perpendicular to the nucleus at greater radial distance in combination with isothermal thickening and increase lamellar separation in untwisted growth. However, beyond a certain distance, inferred to be when thickening has brought branches into the fold surface and stressed the surface, growth becomes twisted and dominant lamellar profiles become slightly S-shaped.[17] The band period is reduced by comparison with faster growth conditions with an abrupt decrease of gradient in the plot against crystallization temperature.

Time is of the essence in achieving optimum surface packing: for sufficiently slow growth, there is more time for the system to explore alternative conformations and reach that of lowest accessible free energy. Experiments with centrally-branched monodisperse alkanes have shown that, at the highest crystallization temperature, lamellae spontaneously adopt cylindrical habits, with all branches on the outer, convex, surface demonstrating that the gain from improved surface packing outweighs the cost of straining the lattice in lamellar interiors.

Banded Growth

Although the relief of surface stress is the reason for banded growth, one still has to account for how such asymmetric growth develops and propagates in what are often symmetric systems. In polyethylene the phenomenology is now known. The essential asymmetry is introduced because the axis of S-profiled lamellae does not lie along the **b** axis, as is to be expected because fold placement is not crystallographic. The S-profiles propagate radially and consistently via arrays of isochiral giant screw dislocations lying to one side of the axis of the S where the shear stress for their formation is highest. The increased compliance of a dislocated lamella allows what is effectively an increment of twist to be given to a new lamella inserted at the dislocation.[18]

Towards Improved Properties

The early appreciation that fold surfaces reduced the longitudinal Young's modulus of polyethylene fibres lead to the introduction of different means to produce materials which had higher modulus but were still tough, such as melt spinning, gel spinning and melt kneading. However, the large extensions employed to attain high molecular extension towards this end produced fibres of, say, ~40 μm diameter. An effective means of cementing such fibres together to fabricate specimens of larger size with substantially maintained properties is hot compaction, in which just sufficient polymer is melted from the periphery of lightly compressed fibres to fill the intervening interstices.[19,20] The later development of two-dimensional compactions has since led to the commercial production of polypropylene sheets with exceptional impact properties.

Fibre Structure and Cold Drawing

In addition to morphological studies helping to guide the development of these and related materials, they have also led to significant gains in understanding the organization of macromolecules in highly-oriented fibres and of the cold drawing process.. Examination of the cross-sections of various advanced polyethylene fibres has shown that they share a common internal substructure, though differing in characteristic detail.[21] After etching one sees craters surrounded by high-melting walls on which later and lower-melting growth has nucleated. Linear voids down the centres of etched craters, akin to those often found in melt-crystallized polymer chip, are indicative of a shortage of material to complete crystallization within the defined volume. The walls themselves are presumably generated when the entangled molecular network sustains the applied stress with their differences of detail, particularly scale, reflecting the specific features of the various processing routes. This is new and pertinent information which must underlie a better understanding of structure/property relationships.

Cold drawing is the name given to deformation of an existing structure as opposed to formation of an elongated texture directly from the melt; it is not restricted to ambient temperature but, e.g. for linear polyethylene, can extend above 100 °C. For a long time it was believed, from small angle scattering perpendicular to the draw direction, that the initial morphology was destroyed at high draw ratios. This is not so. Examination of the cross-sections of polyethylene drawn ~50x has shown that a memory of the initial morphology survives.[22,23] This demonstration that the final properties do relate to those of the starting material opens the door to more-informed processing.

Conclusions

The study of polymer morphology continues to be the principal means of acquiring knowledge and understanding of macromolecular self-organization. Long-standing problems of the nature of melt-crystallized lamellae and spherulitic growth have been resolved, bringing understanding of how characteristic properties such as a broad melting range and spatially-dependent mechanical response stem from inherent systematic variations in the internal microstructure of spherulites. These stem from the distinctive features brought to crystallization by long molecules, namely that they impede each other at the growth front and, for faster growth, form rough basal surfaces prone to subsequent reorganization. Morphological knowledge is an essential accompaniment to the informed development of advanced materials and a full understanding of their structure/property relations.

[1]   D. C. Bassett, "Polymer Spherulites", in: *Encyclopaedia of Materials Science and Technology, 1st Online Update*. K. Buschow, R. Cahn, M. Flemings, E. Kramer, S. Mahajan, P. Veyssiere, Eds., Pergamon, Oxford, 2002.
[2]   D. C. Bassett, *J. Macromol. Sci. B* **2003**, *42*, 229.
[3]   D. C. Bassett, "Polymer Morphology", in: *Encyclopaedia of Polymer Science and Technology, 3rd Edition*, Wiley, New York, 2003.
[4]   A. Keller, *Phil. Mag* **1957**, *2*, 1171.
[5]   F. C. Frank, *Proc. R. Soc. Lond. A* **1964**, *282*, 9.
[6]   F. C. Frank, *Proc. R. Soc. Lond. A* **1970**, *319*, 127.
[7]   S. Y. Lee, R. H. Olley, D. C. Bassett, *J. Mat. Sci.* **2000**, *35*, 5101.
[8]   M. I. Abo el Maaty, *Polym. J.* **1999**, *31*, 778.
[9]   M. I. Abo el Maaty, D.C. Bassett, to be published.
[10]   J. J. Janimak, D. C. Bassett, *Polymer* **1999**, *40*, 459.
[11]   I. L.Hosier, D. C. Bassett, *Polymer J.* **1999**, *31*, 772
[12]   I. L.Hosier, D. C. Bassett, *Polymer* **2000**, *41*, 8801.

[13] I. L.Hosier, D. C. Bassett, *J. Polym. Sci., Polym. Phys. Ed.* **2001**, *39*, 2874.
[14] I. L.Hosier, D. C. Bassett, *Polymer* **2002**, *43*, 307.
[15] M. I. Abo el Maaty, D. C. Bassett, *Polymer* **2001**, *42*, 4957.
[16] M. I. Abo el Maaty, D. C. Bassett, *Polymer* **2001**, *42*, 4965.
[17] M. I. Abo el Maaty, D. C. Bassett, *Polymer* **2002**, *43*, 6541.
[18] D. Patel, D. C. Bassett, *Polymer* **2002**, *43,* 3795.
[19] P. J. Hine, I. M. Ward, R. H. Olley, D. C. Bassett, *J. Mat. Sci.* **1993**, *28*, 316.
[20] R. H. Olley, D. C. Bassett, P. J. Hine, I. M. Ward, *J. Mat. Sci.* **1993**, *28*, 1107.
[21] M. I. Abo el Maaty, R. H. Olley, D. C. Bassett, *J. Mat. Sci.* **1999**, *34*, 1975.
[22] T. Amornsakchai, R. H. Olley, D. C. Bassett, M. O. M. Al-Hussein, A. P. Unwin, I. M. Ward, *Polymer*, **2000**, *41*, 8291
[23] T. Amornsakchai, R. H. Olley, D. C. Bassett, A. P. Unwin, I. M. Ward, *Polymer* **2001**, *42*, 4117.

Crazing and Fracture in Polymers: Micro-Mechanisms and Effect of Molecular Variables

H. H. Kausch,[*1] *J.-L. Halary,*[2] *C. J .G. Plummer*[1]

[1] Institut des Matériaux, Ecole Polytechnique Fédérale de Lausanne (EPFL), Switzerland
E-mail: hans-henning.kausch@epfl.ch

[2] Ecole Supérieure de Physique et Chimie Industrielle de la Ville de Paris (ESPCI), France

Summary: The influence of the primary molecular parameters chain configuration, architecture and molecular weight (MW) on the mode of mechanical breakdown is discussed for two series of (amorphous) thermoplastic polymers, methyl methacrylate glutarimide copolymers and amorphous semi-aromatic polyamides. Structural and dynamic analyses and fracture mechanical methods applied to such adequately chemically modified (glassy) polymers permit us to show and to explain the effect of intrinsic variables on local molecular motions and on the competition between chain scission, disentanglement and segmental slip, which in turn determine the dominant mode of instability and plastic behaviour. Above a critical molecular weight, toughness depends most strongly on the entanglement density; a positive effect of the intensity of sub-T_g relaxations and in-chain cooperative motions on the toughness of these materials is clearly evident.

Keywords: amorphous polyamides; cooperative motions; crazing; entanglements; microdeformation; toughness

Introduction

Good ultimate properties are among the most important prerequisites for the successful use of a polymer material, no matter whether the mechanical, the optical, or some specific functional properties are to be exploited. For this reason the deformation and fracture behaviour of polymers and ways of improving them have been studied intensively.[1-3] The strength and toughness of a thermoplastic construction element depend on the molecular properties of the chosen material, on molecular packing (density, phase structure, micro-morphology) and the way stresses are transmitted between them (through cohesive forces, cross-links or entanglements). The ultimate properties of such polymers depend very strongly on the long-

© 2004 International Union of Pure and Applied Chemistry   DOI: 10.1002/masy.200451003

range transmission of stresses, which requires that the molecular chains form a well entangled (semi-crystalline or amorphous) physical network.

The effect of molecular variables can be studied – or at least estimated – using the following approaches.

1. *Direct experimental investigation* using well-characterised samples. Thus molecular weight (MW) effects on mechanical properties become apparent when comparing mono-disperse samples of different MW. The same applies to the effect of specific molecular groups – such as presence or absence of double bonds, aromatic or bulky substituents, hydrogen bonds etc.

2. *Prediction from theoretical models.* Many physical and viscoelastic properties as well as phase behaviour of (thermoplastic) polymers can be derived from fundamental theories taking into account the chain-like nature of macromolecules, their chemical composition and the connection and interaction between the constitutive groups.[4] A direct correlation of polymer properties with chemical structure has been attempted by the *concept of additive group contributions,* referring to proven and tabulated additive functions such as e.g. the Rao function, the molar refraction or the molar glass transition function.[5] Such an approach has the distinct advantage that it can be applied to new compounds even before they are synthesised. It generally works well with *static, small-strain* properties (bulk modulus, speed of sound, index of refraction), but somewhat less with thermal properties (glass transition temperature T_g). Unfortunately all *ultimate mechanical properties*, which involve large deformations, are not predictable purely from chemical composition.

3. *Computer simulation of material behaviour.* Since about two decades ago molecular dynamics has been modelled by computer simulation. Initially, packing density, its effect on molecular motion and small strain behaviour were studied.[6-8] As an example we mention the simulation of the shear deformation of (atactic) polypropylene, which has revealed discontinuous steps of local reorganisation involving simultaneously more than 100 segments along a chain.[6] The predictive power of this technique is progressing rapidly, now permitting us to *calculate* the elastic, thermal and optical properties of nano-composites and reinforced polymers as well as the compatibility, phase separation behaviour and dynamics of local motion of polymer blends[8] and copolymers.[9]

The approach discussed in this presentation is the experimental investigation of the deformation behaviour of a series of amorphous samples, each of which differs in just one molecular parameter (composition, configuration, chain length). The introduced variations had a direct influence on the transition temperatures between the different elementary damage and deformation mechanisms (chain scission, slip, disentanglement) and thus on the mode of failure and material toughness. The presented results are especially based on two research projects[10-11] in which the Laboratoire de Physicochimie Structurale et Macromoléculaire (PCSM) de l'Ecole Supérieure de Physique et Chimie Industrielle de la Ville de Paris (ESPCI) and the Laboratoire de Polymères de l'Ecole Polytechnique Fédérale de Lausanne (EPFL) have collaborated for the last ten years and which have resulted in numerous publications. [12-18]

The crucial influence of processing conditions, of reinforcing modifications, the presence of defects and of inappropriate design on sample strength should be mentioned, but they will not be discussed in this paper.

Materials

The polymers selected for our investigations were a series of MMA-glutarimide random copolymers (MGLUT, Figure 1) and two series of semi-aromatic polyamides (SAPA, Figure 2 and 3). The principal variable in the MGLUT copolymers is their composition, the glutarimide content, which varied from 36 mol-% (MGLUT36) to 76 mol-%. Within this range the molecular weight M_w increased slightly from 76 to 110 kDa, and the glass transition temperature T_g from 134 ° to 158 °C.[10, 13]

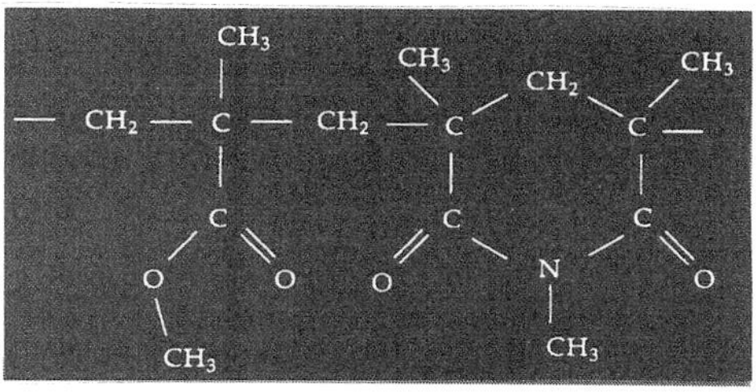

Figure 1. Chemical structure of a methyl methacrylate(MMA) –N-methyl-glutarimide diade.

The second series, referred to as SAPA-A, is based on lactam-12 sequences, terephthalic and/or isophthalic acid residues and 3,3'-diamino 2,2'-dimethyl dicyclohexylmethane residues (Figure 2).

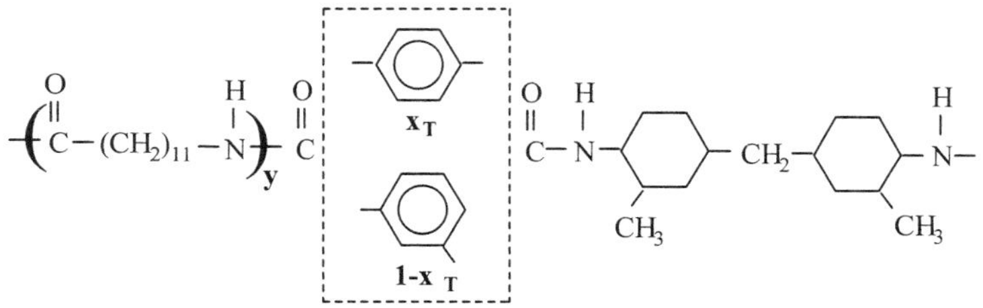

Figure 2. The chemical structure of semi-aromatic polyamides SAPA-A. (See Table 1 for characterisation and designation of samples used.)

Table 1. Molecular and mechanical characterisation of semi-aromatic polyamides SAPA-A [data from 11, 16, 17].

Sample (designation)	y	x_T	M_w $g \cdot mol^{-1}$	T_α^* °C	ρ^* $kg \cdot m^{-3}$	M_e $g \cdot mol^{-1}$	$10^{-26} \nu_e$ m^{-3}	K_{Ic}^* $MPa \cdot m^{1/2}$
A-1.8I	1.8	0	22 000	130	1 042	2 700	2.3	2.3
A-1.8T(23)	1.8	1	23 000	137	1 042	3 000	2.1	2.45
A-1I(26)	1	0	26 000	161	1 055	2 800	2.25	2.35
A-1T$_{0.7}$I$_{0.3}$(23)	1	0.7	23 000	171	1 057	3 100	2.0	2.45
A-1T$_{0.7}$I$_{0.3}$(32)	1	0.7	32 000	171	1 057	3 050	2.0	2.5

$^*T_\alpha$ = temperature at maximum of dynamic loss modulus at 1 Hz; ρ = density at 25 C,

K_{Ic} = critical stress intensity factor at 20 °C

Here the principal molecular variables are the molecular weight M_w, the relative amounts of the lactam-12 sequences (y) and the configuration of the phenyl ring linkages (x_T=0 designates the meta- and x_T=1 the para-position), which influence entanglement molecular weight M_e and entanglement density ν_e.

The third series, referred to as SAPA-R, was based on 2-methyl 1,5-pentanediamine and terephthalic and/or isophthalic acid (Figure 3). The configuration of the phenyl ring linkages and the small mobility of the pentanediamine group will have to be considered.

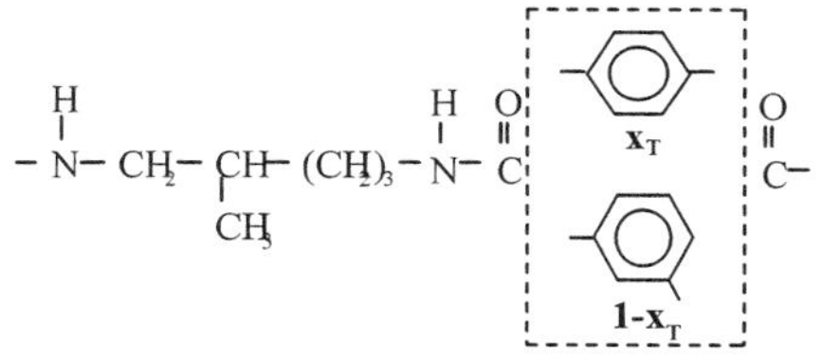

Figure 3.   Chemical structure of SAPA-R (see Table 2 for characterisation and designation of used samples).

Table 2. Molecular and mechanical characterisation of semi-aromatic polyamides SAPA-R data from ref.[11, 16, 17]

Sample	x_T	M_w	T_α	ρ	M_e	$10^{-26} \nu_e$	K_{1c}^*
(designation)		$g \cdot mol^{-1}$	°C	$kg \cdot m^{-3}$	$g \cdot mol^{-1}$	m^{-3}	$MPa.m^{1/2}$
R-I	0	18 000	141	1 194	2 750	2.6	3.8
$R-T_{0.5}I_{0.5}$	0.5	23 000	145	1 196	2 900	2.5	4.0
$R-T_{0.7}I_{0.3}$	0.7	22 000	147	1 196	2 950	2.45	3.8

Micro-deformation Mechanisms

The members of all three series were amorphous, with glass transition temperatures above 130 °C. When straining thin films of these materials they all showed similar micro-deformation mechanisms involving scission crazing at low temperatures, crazing and formation of deformation zones in different forms in a large temperature range around and above 50 °C disentanglement crazing at more elevated temperatures and homogeneous deformation at T close to T_g. However, the exact temperature-deformation mechanism map and especially the transition temperatures, at which one form changes into another, depend on the molecular composition. For MGLUT76, representative micrographs of these micro-deformation mechanisms are shown in Figure 4. The temperature ranges where these mechanisms were observed are indicated in Figure 5.

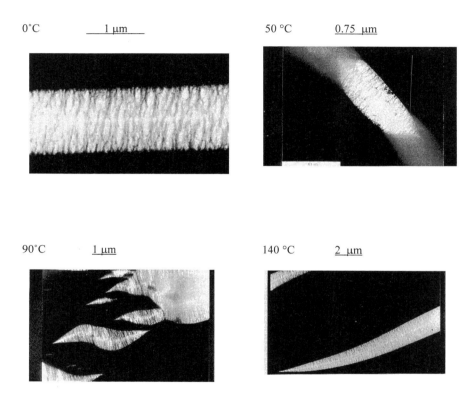

Figure 4.   Micro-deformation mechanisms observed in thin films of MGLUT76 strained at a rate of $2 \cdot 10^{-3} s^{-1}$. Chain scission crazing at 0°C, shear blunting at 50 °C multiple crazing at craze tips (90 °C) and disentanglement crazing (140 °C. The glass transition occurs at 158 °C.

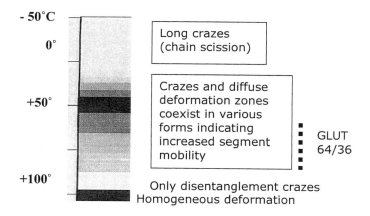

Figure 5. Temperature-deformation mechanism map of MGLUT76. Dark zones indicate notable shear deformation. For comparison, the temperature range where crazes and diffuse deformation zones (DZ) are observed together is also shown for MGLUT36 (dashed line). Note that in the MMA-rich MGLUT36 Dzs do not appear below temperatures of around 50 °C, whereas they are observed at 20 °C in MGLUT76 after ref. [10, 12]

The observed phenomena can be explained using the current theory on craze initiation and growth. [1, 3, 13, 17-21] Crazes grow by transformation of a glassy matrix material into fibrillar craze matter, which necessarily involves a loss of entanglements. There are three mechanisms by which the number of entanglements can be reduced: chain scission, forced reptation and slippage against van der Waals forces. These three mechanisms have quite different kinetics. At low temperatures (in the MGLUT system at $T_g - T > 100$ K) the entangled chains show little mobility; they are held so tightly that they will rather break than slip. Chain scission requires very high stresses (to break a C-C bond a force of 4 nN is required which corresponds to a stress acting on the chain cross-section of about 6 GPa [1]). Such stresses are only found in regions of high stress concentration (at craze tips and within the process zones at the fibril-bulk interface). This stress geometry gives rise to (long and thin) crazes of high aspect ratio. With increasing chain mobility overstressed chains are able to transfer part of the load and to escape scission temporarily. Craze fibrils become stronger and longer, the rate of deformation of craze tips and at the craze-bulk interface decreases, which favours further orientation and the formation of (diffuse) shear deformation zones (DZ). In the intermediate temperature region, therefore, mixed forms of plastic instability are found (Figure 4 and 5). Above 80 °C chain mobility is sufficiently high so as to favour disentanglement by forced reptation and crazing becomes the dominant deformation mechanism. In MGLUT76, at some 15 to 30 °C below T_g, it is the only active mechanism as evidenced by the appearance of the sharp tipped crazes (Figure 4). On the other hand, very close to T_g (158 °C), stress relaxation is too high to permit any form of localised deformation; the sample deforms homogeneously.[10, 13]

Like many other (amorphous) polymers the materials SAPA-A and -R show basically the same sequence of deformation mechanisms. However, the transition temperatures and the relative importance of the different mechanisms depend on molecular composition and influence notably the toughness of these materials, thus making them excellent candidates for a systematic study of the effect of molecular variables on ultimate mechanical properties.[11, 16, 17]

Mechanical Testing

The toughness of a given material is characterised by the critical stress intensity factor K_{Ic}. Plain strain fracture tests were performed in mode I on standard three-point bending samples (Figure 6), whose critical dimensions satisfy the criteria of the *Williams protocol*.[22]

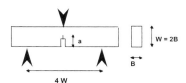

Figure 6.   Three-point bending specimen for fracture toughness testing (*a* crack length, B specimen thickness, W width).

For good reproducibility the machined notch had been sharpened by a pre-crack, introduced with the help of a specially designed falling weight apparatus, each time using a fresh razor blade. Both the length of the pre-crack and the sharpness of the crack tip were checked for quality prior to testing by optical microscopy.[11, 16, 17] The samples were then loaded in an MTS 810 testing machine at the desired temperatures at a constant crosshead displacement rate of 1 mm/min. The K_{Ic} values were calculated from the expression:

$$K_{Ic} = f(a/W)\frac{P_{max}}{BW^{1/2}} \tag{1}$$

where f(a/W) is a standard correction function [21, 22] and P_{max} the maximum recorded load. It is useful to remember that the tip of the pre-crack consists of a craze, whose stability and tendency to transform into a plastic zone determine the fracture behaviour; in the case of brittle fracture P_{max} designates the load, where unstable crack propagation begins, in the case of ductile fracture it is the load, where the increase due to continued straining of the cracked sample and the decrease due to stress relaxation and crack growth are in equilibrium. The critical energy release rate, often simply called *fracture energy* G_{Ic}, is obtained from

$$G_{Ic} = \frac{U_i}{BW\phi} \tag{2}$$

where U_i is the area under the load-displacement curve integrated up to P_{max}, Φ is another tabulated correction function.[21, 22]

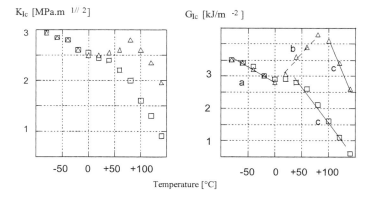

Figure 7.   Toughness (K_{Ic}) and fracture energy G_{Ic} of two SAPA A-1$T_{0.7}$I $_{0.3}$ materials having molecular weights of 23 ($\square$) and 32 kDa ($\triangle$) respectively (from ref.[11, 16]). Three different regions (a, b, c) are observed (see text for discussion).

Representative data of toughness development as a function of temperature, obtained for two semi-aromatic polyamides SAPA A-1$T_{0.7}$I $_{0.3}$ of different molecular weight are shown in Figure 7.

Attention is drawn to the fact that the fracture resistance of the low MW material decreases at slowly first, then more rapidly, The high MW material, however, shows three different regions of slightly decreasing (a), then increasing (b) and finally strongly decreasing (c) fracture energy. This behaviour will be analysed in the next section.

Effect of Chain Length and Molecular Composition on Toughness

At low temperature (T < 0 °C) the two SAPA A-1$T_{0.7}$I $_{0.3}$ materials break in a brittle manner by unstable extension of a craze, with craze growth and breakdown controlled by chain scission. Figure 7 reveals that in this temperature region there is no effect of molecular weight. It must be concluded that the overstressed chain segments of even the low MW material ($M_w/M_e = 7.4$) are so tightly anchored that they will rather break than slip. With increasing temperature the stress for craze initiation by chain scission decreases slightly (region a) and with it P_{max} and K_{Ic}.[16, 21] With further increasing temperature chains become more mobile, the transformation of matrix material into fibrillar matter in the process zone occurs at an ever-decreasing stress. As before, it is argued that an increasing number of chains escape chain scission. In the low MW material the chains are too short, however, to stabilize the process zone and the fibrils will fail eventually, $K_{Ic}(T)$ decreases (region c). The increased mobility has a different effect in the high MW material, since the increasing number of chains which escape chain scission lead to longer and stronger fibrils, P_{max} and thus $K_{Ic}(T)$ increases within this region (b). At still higher temperatures, however, (region c, at T_g - T < 70 K) disentanglement becomes the dominant mechanism, even of the longer chains and $K_{Ic}(T)$ decreases (region c).

Besides chain length the intrinsic molecular variable of strongest influence on toughness is entanglement density. The available $K_{Ic}(\nu_e)$ data compiled in Figure 8 show that, on average, the data correspond to the good correlation between K_{Ic} and ν_e first proposed by Wu.[23] However, the three series studied, MGLUT, and SAPA-A and –R, show systematic deviations, which can be ascribed to particular structural characteristics of the involved molecules.

Within the MGLUT series K_{Ic} increases with increasing glutarimide content although the entanglement density remains constant. Tézé[10] and Tordjeman et al.[24] have related this increase of K_{Ic} to a strengthening of the cooperative character of the β-relaxation motion with increasing glutarimide content. The improved cooperativity leads to an increase of the β-relaxation peak and – as we have seen for thin films (Figure 5) – to a lowering of the transition temperature from scission crazing to formation of diffuse deformation zones. In bulk samples at room temperature a similar shift from unstable to stable fracture occurs with increasing glutarimide content, accompanied by a noticeable increase in toughness as shown in Figure 8 (data from Tézé [10]).

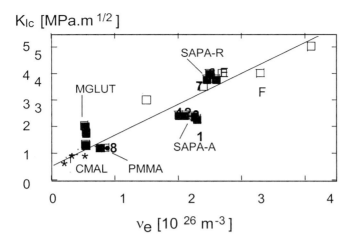

Figure 8. Compilation of room temperature toughness values as a function of entanglement density for different amorphous polymers. □ data from Wu,[23] ■,✱ data from Tézé,[10] Brûlé et al. [11, 16] (see text for discussion).

The series SAPA-A and –R are distinguished by their much higher entanglement density as compared to MGLUT; as is to be expected, R is tougher than A. Within these series, however, the opposite effect is observed, little (or negative) variation of K_{Ic} with v_e. Looking at the molecular structure (Table 2) one notes that it is systematically the terephthalic chains, which have the higher K_{Ic} despite their lower entanglement density. From dynamic mechanical analysis (DMA) one knows that the terephthalic samples have a much stronger β-relaxation peak. As shown by Beaume et al. by ^{13}C-solid state NMR[25] the cooperative character of the β-relaxation motion is strengthened by the capacity of the para-disubstituted phenyl rings to execute π-flip motions. For evident steric reasons the isophthalic samples cannot undergo such π-flip motions.

The fact that a constructive interaction (*cooperativity*) between the molecular motions of neighbouring groups within a chain intensifies the β-relaxation peak has been shown for many polymers.[1, 10, 11, 16-18, 24-27] Plummer et al. have studied bisphenol-A polycarbonate derivatives composed of different alternating blocks of varying lengths.[28] From the obtained DMA spectra they conclude that the β-relaxation involves an in-chain cooperative motion extending across 6-9 repeat units and that this motion is also influential in activating the disentanglement crazing at elevated temperatures.[27] It is this capacity of the chain backbone to rapidly relax axial stresses which also is responsible for the positive correlation between the position of the maximum of the β-peak and toughness, which exists for the majority of amorphous or semi-crystalline polymers.[26]

Conclusions

Structural and dynamic analysis as well as fracture mechanical methods applied to adequately chemically modified (glassy) polymers permit us to show and to explain the effect of the principal intrinsic variables configuration, chain length and entanglement density. The competition between the elementary deformation mechanisms chain scission, segmental slip and disentanglement determines the mode of fracture and the toughness of amorphous polymers. The dominant mode changes with temperature and is strongly influenced by the intrinsic variables. For long chains $(M_w > 9\ M_e)$ toughness depends most strongly on entanglement density and the intensity of sub-T_g relaxations, with in-chain cooperative motions playing an important role. The present studies on the internal toughening parameters of a polymer obtain a particular significance in view of the recent investigations of Grein et al.[28, 29] These authors have shown (for rubber-toughened polypropylene) that particle modification facilitates craze and crack *initiation*, but the essential energy dispersion occurs during *propagation*, which requires the relaxation of the triaxial state of stress by local *matrix* deformation mechanisms.

[1]   H. H. Kausch, *"Polymer Fracture"*, 2nd edition, Springer, Heidelberg-Berlin 1987.
[2]   A. J. Kinloch, R. J.Young, *"Fracture Behaviour of Polymers"*, Appl. Sci. Publ., London-New York 1983.
[3]   H. H. Kausch, C. J. Plummer, N. Heymans, P. Decroly, *"Matériaux Polymères: Propriétés Mécaniques et Physiques"*, Presses Polytechniques et Universitaires Romandes, Lausanne 2001.
[4]   R. N. Haward, ed. *"The Physics of Glassy Polymeres"*, Appl. Sci. Publ., London 1973.
[5]   D. W. van Krevelen, *"Properties of Polymers"*, Elsevier, New York 1990.
[6]   P. H.Mott, A. S. Argon, U. W. Suter, *Phil. Mag. A* **1993**, *67*, 931.
[7]   W. L. Mattice, U. W. Suter, *"Conformational Theory of Large Molecules"*, John Wiley Sons, New York **1994.**
[8]   W. H. Jo, J. S.Yang, *Adv. in Polymer Sci.* **2002**, *156*, 1.

30

[9]   B. Lousteaux, PhD Thesis, University Pierre and Marie Curie, Paris, 2001.
[10] L. Tézé, PhD Thesis, University Pierre and Marie Curie, Paris, Nov. 10, 1995.
[11] B. Brûlé, PhD Thesis, University Pierre and Marie Curie, Paris, Sept. 15, 1999.
[12] H. H. Kausch, C. J. G. Plummer, L. Tézé, *Physica Scripta*, **T55**, *216*, 1994.
[13] C. J. G. Plummer, H. H. Kausch, L. Tézé, J. L. Halary, L. Monnerie, *Polymer* **1996**, *37*, 4299.
[14] L. Tézé, J. L. Halary, L. Monnerie, L. Canova, *Polymer* **1998**, *40*, 971.
[15] B. Brûlé, J. L. Halary, L. Monnerie, *Polymer* **2001**, *42*, 9073.
[16] B. Brulé, L. Monnerie, J. L. Halary, ESIS TC4 2003.
[17] B. Brûlé, H. H. Kausch, L. Monnerie, C. J. G. Plummer, J. L. Halary, *Polymer* **2003**, *44*, 1181.
[18] J .L. Halary, L. Monnerie, Int. Conf. on Deformation, Yield and Fracture, Cambridge, April 2003.
[19] H. H. Kausch Ed., *Advances in Polymer Science* **1990**, 91/92.
[20] G. H.Michler, *Kunststoff- Mikromechanik*, Carl Hanser Verlag, München/Wien 1992.
[21] C. J. G. Plummer, *Current Trends in Polymer Sci.* **1997**, *2*, 125.
[22] ISO 13586-1 Determination of fracture toughness (G_{Ic} and K_{Ic}) for plastics. A LEFM approach.
[23] S. Wu, *J. Polym. Sci., Polym. Phys. Ed.* **1989**, *27*, 723.
[24] P. Tordjeman, L. Tézé, J. L. Halary, L. Monnerie, *Polym. Eng. Sci.*, **1997**, *37*, 1621.
[25] F. Beaume, B. Brûlé, J. L. Halary, F. Lauprêtre, L. Monnerie, *Polymer* **2000**, *41*, 5451.
[26] F. Ramsteiner, *Kunststoffe* **1983**, *73*, 148.
[27] C. J. G. Plummer, C. L. Soles, C. Xiao, J. Wu, H. H. Kausch, A. F. Yee, *Macromolecules* **1995**, *28*, 7157.
[28] C. Grein, PhD Thesis No. 2341, Ecole Polytechnique Fédérale de Lausanne, 2001.
[29] C. Grein, C. J. G. Plummer, H. H. Kausch, Y. Germain, Ph. Beguelin, *Polymer* **2002**, 3279.

Environmental Stress Cracking of Polymers Monitored by Fatigue Crack Growth Experiments

Volker Altstaedt,[*1] *Sven Keiter,*[1] *Michael Renner,*[1] *Alois Schlarb*[2]

[1] Technical University Hamburg-Harburg, Polymer Engineering, Hamburg, Germany

[2] B. Braun Medical AG, Escholzmatt, Switzerland

Summary: This article describes fatigue crack growth experiments to investigate the degradation of the durability of polymers due to fluid environments. The degrading effect of media causing stress cracking can be observed on the fracture surfaces of tested samples by scanning electron microscopy. Strategies to improve environmental stress cracking like changes in molecular weight, orientation, toughening with rubber particles of different sizes are discussed. Fatigue crack growth experiments can be employed as a very fast and effective screening method.

Keywords: fatigue crack growth; mechanical properties

Introduction

Fatigue crack propagation (FCP) experiments can be employed as a fast and effective screening method for determining long-term mechanical properties of polymers. Advantages of this method, like the need for a very small quantity of test material (<10 g), the broad range of fatigue crack propagation rates (from 10^{-2} mm/cycle to 10^{-7} mm/cycle) measured within one specimen and a well defined stress state at the crack tip favor the use of FCP experiments instead of traditional S-N curves [1]. These experiments can also be conducted in the presence of a specific environment and are

   DOI: 10.1002/masy.200451004

able to provide valuable information about environmental effects on crack propagation.

If the FCP experiment is carried out in the presence of a critical fluid environment, a more or less progressive degradation of mechanical properties (embrittlement), depending on the polymer, can be observed. This degradation is caused by enhanced disentanglement and chain scission of the molecules affected by a solving liquid. The process is called environmental stress cracking (ESC). ESC is defined as the simultaneous action of stress and contact with specific fluids. Approximately 15 % of all failures of polymer components are due to ESC. Since prediction in many cases is exceedingly difficult, assessment   through suitable laboratory tests becomes important [2]. Standard tests for environmental stress cracking of polymers are the *Ball or Impression Method* (ISO 4600), the *Bent Strip Method* (ISO 4599) and the *Constant Tensile Stress Method* (ISO 6252).

In these standards either a constant static load or a constant deformation is applied. In the case of the bent strip method and the impression method the residual strength is quantified through a succeeding impact or tensile test. This, however, does not imply that the stress state at the crack tip in the moment of failure is well defined. Moreover, the common standardized test methods do not consider a well defined stress state within the test specimen. This could be achieved by applying the principles of linear elastic fracture mechanics to ESC test procedures in combination with a compact tension (CT) specimen. In contrast to tensile or impact tests, the stress state at the crack tip of a CT-specimen is well defined. A specific fluid is able to penetrate to the front of the crack tip. Under cyclic loading, the propagation of a fatigue crack through the bulk material of a CT-specimen is affected by the presence of the fluid. From the dynamics of crack propagation, information about the interaction between the crack tip, designated as a microscopically small probe, the fluid environment and the ESC sensitivity of a specific polymer, polymer blend or polymer composite, can be achieved.

Fatigue crack growth

All fatigue failures in polymers or polymer composites involve one phase in which a defect zone such as a craze or microcrack initiates, followed by a propagation phase to final fracture. Based on the assumption that the fatigue lifetime is determined by the propagation phase, a preexisting flaw is assumed. The stress state at the tip of the crack is defined by the stress intensity factor range ΔK. For the case of fatigue, Paris [3] showed that a linear relationship predicted by a simple power law for a double logarithmic scale exists between the FCP rate da/dN and the applied ΔK (Figure 1).

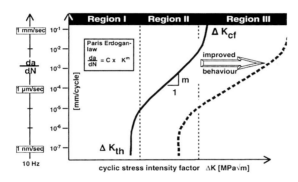

Figure 1.   Scheme of fatigue crack propagation diagram.

The linear dependence is frequently observed only over an intermediate range of growth rates. When investigating a wide range of da/dN, deviations from this linear behavior may be observed, as illustrated schematically in Figure 1. That is, FCP-rates decrease rapidly to vanishingly small values as ΔK approaches the threshold value ΔK_{th}. This ΔK level defines a design criterion that is analogous to the fatigue limit determined from traditional S-N curves [4,5]. FCP rates increase markedly as ΔK approaches ΔK$_c$, at which unstable fracture occurs within one loading cycle. From the standpoint of evaluating a materials fatigue resistance, any decrease in FCP rates at a given value of ΔK or, alternately, any increase in ΔK to drive a crack at a given speed is, of course, beneficial.

As shown in Figure 2, the experiments can be conducted also under constant ΔK conditions. In this case, specific influences of a medium could be detected by a change in the crack propagation rate as a function of exposure time. For metallic materials, fatigue crack growth experiments under environmental conditions are described in ASTM E 647, but for polymers, no standard procedure exists until now.

Environmental stress cracking (ESC) of polymers is a phenomenon which has been researched over a period of more than 40 years. The phenomenon involves so many influential variables that the behavior cannot be predicted with sufficient accuracy. The only alternative is testing. It is for this reason that any advances in test methods are important, to make research more efficient and effective.

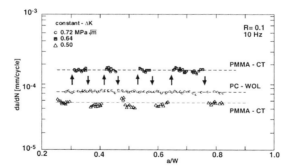

Figure 2. Experiments conducted under constant K [6] conditions.

Experimental

Materials

Amorphous thermoplastics are sensitive to ESC. In particular, polystyrene (PS) is well known to be sensitive to fluid permeation causing crazing [2]. Commercial PS polymers were selected for the investigation of the effect of molecular weight and various types of rubber modification on the ESC behavior under fatigue crack growth conditions. The materials were kindly supplied by BASF AG, Ludwigshafen, Germany and are listed in Table 1.

Two different polycarbonates were used for the investigations regarding lipid resistance. The specimens were kindly supplied by B. Braun Medical AG, Switzerland.

All polymers were processed by injection molding under standard conditions and tested considering the injection molding direction.

Test methods

Compact-type (CT) specimens were cut from rectangular injection moulded plates with 4 mm nominal thickness and precracked by a razor blade. All fatigue crack growth tests were run on a Schenck PSA® servohydraulic test system at a frequency of 10 Hz under sinussoidal loading in tension-tension with a minimum to maximum load ratio (R) of 0.1 at room temperature. The tests were run under ΔK control with a software designed by Fracture Technology Associates, Inc. A ΔK-decreasing portion and a ΔK-increasing portion were measured separately for two different specimens of the same material and combined to one FCP-diagram. The crack length was monitored by a compliance technique, as published by Saxena & Hudak [7].

The tests under environmental conditions were carried out by applying the critical fluid through a soaked sponge, which was fixed on both sides of the specimen as an unlimited source. According to this procedure the crack was always covered with the soaked sponge. For the test with polystyrene (PS) and a commercial sunflower oil (tradename Livio®) were used. The tests with Polycarbonate (PC) were performed with a fat emulsion (Lipofundin® MTC 20 %) for parenteral nutrition.

In this investigation all tests were run within load amplitude limits under ΔK control. The fatigue behavior was expected to deteriorate because, under the softening effect of a liquid medium, the samples would be strained more extensively for each loading cycle.

Table 1.   Investigated materials

material	property
PS 148H	polystyrene - M_w 238,000 g/mol
PS 168N	polystyrene - M_w 354,000 g/mol
PS 486M	polystyrene impact modified by small PB particles
PS 2710	polystyrene impact modified by large PB particles

Fractography

Fracture surfaces were studied with a field emission electron microscope LEO 32. The microscope was operated at an accelerating voltage of 0.5-1 kV. Because of the low accelerating voltage no gold coating of the specimen was necessary and the specimens could be investigated directly after fracture of the specimen in the fatigue experiment.

Results and Discussion

Effect of molecular weight

Physical properties of polymers are strongly dependent on the average molecular weight M_W and sequence distribution M_W/M_N, because the presence of molecular entanglements can significantly affect the mechanical behavior. As previously shown by Altstädt [8] for different commercial PS systems prepared by free radical polymerization and by anionic polymerization, the number average M_N of the molecular weight corresponds well to an increase of the fatigue crack growth behavior. Further experiments were conducted in this study to explore the effect of environmental ageing.

The effect of molecular weight on the ESC behavior was investigated for two different PS systems. These were tested with and without an oil environment. As shown in Figures 3 – 4, both PS systems exhibit a severe decrease in fatigue crack growth resistance when tested under oil environment. In both cases, ΔK for a given fatigue crack growth rate is significantly reduced and the slope of the curve is increased. Simultaneously, a decrease in ΔK_{th} is observed. The comparison of both PS systems clearly shows an improved ESC resistance under fatigue loading of the PS with the higher molecular weight. This is reflected in a smaller slope of the linear portion of the FCP-diagram and a higher ΔK_C and ΔK_{th} for PS 168N.

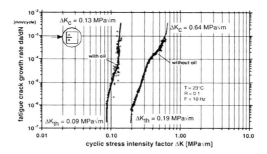

Figure 3.   FCP with and without stress cracking media.

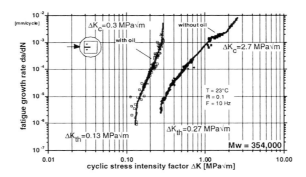

Figure 4.   FCP with and without stress cracking media.

Effects of orientation

To investigate possible orientations of the molecules due to the injection moulding process on the ESC resistance under fatigue loading conditions, samples of polystyrene were tested parallel and perpendicular to the injection moulding direction.   A specimen prepared from granules by compression molding, which should have a minimum of orientation, was included in the investigation as a reference.   To minimize the effect of a plastic deformation possibly induced by precracking with the razor blade, the precrack was extended for a minimum 2 mm by fatigue loading before starting the experiment.   In some cases it was difficult to carry out the test, because the crack grew parallel to the injection direction or an additional crack was initiated on the fixing holes of the samples.

As shown in Figure 5 the FCP-diagram of the specimens tested perpendicular to the injection moulding direction is shifted significantly to higher ΔK values compared to the specimens tested parallel.   While the reference specimens processed by compression molding behave similar to those with parallel orientation.   Obviously it is easier for the fatigue crack to propagate in the direction of the oriented entanglement network.   This can be explained by the fact that crack propagation in PS is accompanied by crazing.   Molecules which are already stretched in one direction are losing the ability to fibrillate in the other direction. By this, the probability for chain scission is increasing and the breakdown of the material drawn into the process zone occurs at lower ΔK values.

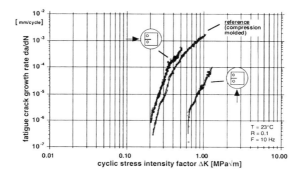

Figure 5. Molecule orientation effects on crack propagation.

If polystyrene is tested perpendicular to the injection molding direction and a stress cracking media is present at the same time, an additional embrittlement can be observed by the steep increase in the FCP-curve, which makes it almost impossible to measure the dynamic of crack propagation in a broad range of crack propagation rates as usual (Figure 6).

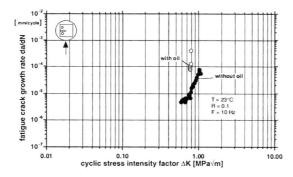

Figure 6.   Injection moulded samples tested perpendicular to the injection direction.

Toughness modification by rubber particles

A common strategy to improve the mechanical behavior of thermoplastics is to incorporate uniformly distributed rubber particles in the polystyrene matrix. In the case of HIPS (high impact polystyrene) styrene-butadiene block copolymers are used. This strategy is also applied to improve the environmental stress crack resistance of polystyrene.

For this investigation two commercial PS grades with different average diameters of the rubber particles were selected (PS 486M: $\varnothing \approx 2$ μm, PS 2710: $\varnothing \approx 5$ μm).

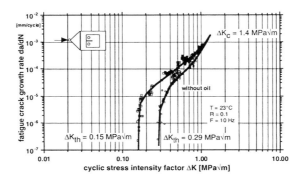

Figure 7.   Effect of small styrene-butadiene rubber particles incorporated in PS.

The systems were tested with and without exposure to vegetable oil. The corresponding FCP-diagrams are shown in Figures 7 - 8.

The FCP behavior of the two rubber-modified polystyrenes is comparable in the threshold region and in region III (Figure 1) at high propagation rates. A significant difference is visible in region II of intermediate crack propagation rates 10^{-5} to 10^{-4} mm/cycle. In particular, PS with the larger rubber particles is more effective in decreasing the slope of the crack growth curve.

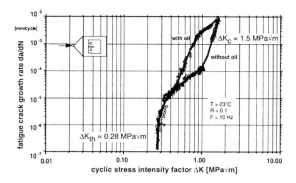

Figure 8.   Effect of large styrene-butadiene rubber particles incorporated in PS.

Compared to the unmodified PS 148H and PS 168N (Figures 3 - 4) the observed ranking in terms of FCP behavior is PS 148H < PS 486M = PS 2710 < PS 168N.   This ranking is changed completely if the tests are conducted in the presence of vegetable oil: PS 148H << PS 168N < PS 486M < PS 2710. Obviously rubber toughening is more efficient for the improvement of the ESC resistance under fatigue crack growth conditions than the increase in molecular weight corresponding to PS 148H and PS 168N.   A remarkable difference between the two rubber modified systems is visible in the threshold region.   The ΔK_{th} value of PS 2710 is not affected by the presence of the medium while ΔK_{th} of PS 486M is reduced by a factor of two.

The observation of the fracture surfaces in Figures 9 - 12 reveals the different sizes of the rubber particle of the two PS systems (Figures 9 - 11).   Tested in air, in both cases the crack propagates through the rubber particles which is only possible in the case of good interfacial bonding.   Tested in vegetable oil, the fracture surface appears to be very smooth.   In the case of PS 486M, some marks from the underlying morphology are visible.

It is generally accepted that rubber particles act as stress concentrators, initiate crazes as well as participate in their termination. The contribution of toughness depends on the concentration of the rubber, the particle size and the interfacial bonding. In the case of fatigue crack, growth of small particles is more efficient at low crack propagation rates because the plastic zone size is smaller, while larger particles are more efficient at higher propagation rates and the corresponding larger plastic zone diameter.

Under the influence of a stress cracking medium it seems that in the case of large rubber particles small crazes may easily develop. As a consequence the diffusion rate of oil is reduced, because of the increased fibril density. Small rubber particles show less intensive crazing and the crazes are easier overloaded because of a higher effective opening of the crack tip. In the latter case the oil can penetrate easier through the crazes. This effect is stronger in region I of the FCP curve as compared to region II (Figure 1). At high crack propagation rates, it seems to be possible, that the oil is not able to penetrate fast enough to the crack tip which explains the same ΔK_C values of the two rubber modified systems.

It should be mentioned at this point, that our interest here is rather in the effect of critical fluids on fatigue crack propagation. The total fatigue lifetime of unnotched specimens is determined by a initiation and a propagation phases. The initiation phase could be significantly effected by the presence of rubber particles [9]. The effect of critical fluids on the initiation phase has to be studied separately.

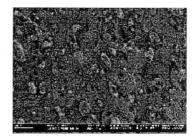

Figure 9.   FCP fracture surface of PS 486 486M tested under air environment.

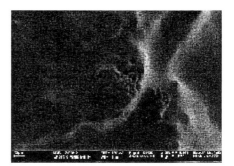

Figure 10. FCP fracture surface of PS 486 486M tested under oil environment.

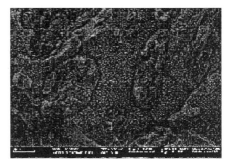

Figure 11. FCP fracture surface of PS 2710 tested under air environment.

44

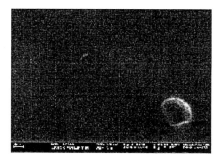

Figure 12. FCP fracture surface of PS 2710 tested under oil environment.

Lipid resistance of medical devices

For medical devices, in particular for components of infusion systems, transparent polymers are indispensable. Because of the multitude of drugs possibly flowing through these systems, the need for disinfection of the devices and the possibility to connect various components by force locking, the ESC resistance plays a decisive roll in the selection of a suitable material. Figure 13 shows a three-way stop cock as a important component of such a system.

In principle, the ESC resistance can be controlled within limitations, which are mostly given by the viscosity for processing, by the molecular weight and molecular weight distribution of the polymer. Because of this, polycarbonate with a molecular weight of above 30,000 g/mol is used for this application.

Specially in the case of parenteral nutrition with lipid-containing emulsions the occurrence of stress cracking by the intravenous infusion in three-way stop cocks made of polycarbonate can not be completely excluded.

Figure 13. Three-way Stop cock as a component for medical infusion systems.

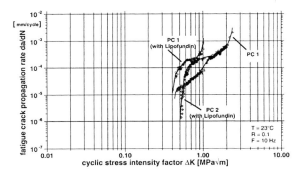

Figure 14. FCP with and without stress cracking media in polycarbonate PC1 (reference)
and PC2 (improved).

A significant improvement in the administration of a lipid-containing emulsion was achieved with a
special additive to polycarbonate. As shown in Figure 14, the better behavior of the new
polycarbonate PC2, proved in practice, could also by verifyed by fracture mechanical fatigue crack
growth experiments. In the presence of the fat emulsion the higher lipid resistant PC2 shows a
higher fatigue threshold value $\Delta K_{th,}$ as well as an improvement by a factor of two for ΔK_c .

Conclusions

Fatigue crack propagation (FCP) experiments can be employed as a fast and effective screening method for the evaluation of environmental effects on crack propagation. As shown for PS, molecular weight, molecular orientation and rubber toughening play an important role in the ESC behavior.

Acknowledgement

We thank Dr. F. Ramsteiner and Dr. W. Loth from BASF AG, Polymer Research Division, Ludwigshafen, Germany for the helpful discussions.

[1]   Altstädt V. & Loth W. & Schlarb A. 1995, *Comparison of Fatigue Test Methods for Research and Development of Polymers and Polymer Composite*s; Proceeding of the International Conference on Progress in Durability of Composite Systems. Brussels, Belgium, 16-21.7.1995.

[2]   Kambour, R.P. & Gruner C.L. & Romagosa E.E. 1973, J. Polymer. Sci., Polym. Phys., Vol.11, p.1879

[3]   Paris, P. C. 1964, Proceedings of the 10th. SagamorConference. Syracuse Univ. Press, NY, p. 107.

[4]   Hertzberg R.W. & Manson J.A. 1990, F*atigue of Engineering Plastics* New York: Academic Press.

[5]   Janzen, W. & G.W. Ehrenstein 1991. *Bemessungsgrenzen von glasfaserverstärktem PBT bei schwingender Beanspruchung.* Kunststoffe 81-3, p. 231.

[6]   Lang, R.W. 1984, *Applicability of Linear Elastic Fracture Mechanics to Fatigue in Polymer and Short - Fiber Composites;* Ph.D. Dissertation, Lehigh University, Bethlehem, PA, USA.

[7]   Saxena A. & Hudak S.J. & Donald J. K. & Schmidt D.W. 1978, J. Test Eval. 6   p 167.

[8]   Altstädt V. 1997, *Fatigue Crack Propagation in Homopolymers and Blends with High and Low Interphase Strength*; Proceedings "European Conference on Macromolecular Physics - Surfaces and Interfaces in Polymers and Composites", Lausanne, June 1st-6th 1997.

[9]   Sauer, J. A. & Chen C. C., *Crazing and Fatigue Behavior in One- and Two-Phase Glassy Polymers*; Advances in Polymer Science, 52/53, Editor: H. H. Kausch.

[10]   Bubeck 1981, Polym. Eng. Sci. 21, p 624.

Macromol. Symp. **2004**, *214*, 47-71

Toughness Enhancement of Nanostructured Amorphous and Semicrystalline Polymers

Goerg H. Michler,* Rameshwar Adhikari, Sven Henning

Institute of Materials Science, Martin Luther University Halle-Wittenberg,
D-06099 Halle/Saale, Germany
E-Mail: michler@iw.uni-halle.de

Summary: An overview is given of different micromechanical deformation processes leading to an enhancement of toughness in heterophase polymers. The well-known mechanism of rubber or particle toughening of semicrystalline polymers was studied in HDPE and PP blends. In particular, the micromechanical processes in the semicrystalline polymer strands between modifier particles were investigated in detail, revealing processes of separation, yielding, breaking and twisting of lamellae. These processes are compared with lamellae forming amorphous SBS block copolymers with alternating soft (polybutadiene) and hard (polystyrene) layers. Depending on the deformation direction, the mechanism of thin layer yielding or chevron formation appears. In both polymeric systems, the initial stage of deformation is characterized by a plastic yielding of the soft phase with a reorganization of the hard (glassy or crystalline) lamellae. The second stage is determined by the alignment of the hard phase towards the deformation direction and the plastic yielding. Detailed comparison of these similar mechanisms in very different polymers with similar nanostructured morphology should help to improve toughening of amorphous as well as semicrystalline polymers.

Keywords: electron microscopy; micromechanical mechanisms; morphology; polyolefins; SBS block copolymers: thermoplastic polymers; toughness enhancement

Introduction

Toughness of polymers is an important property for many practical applications. In this context, the term 'toughness' denotes the absorption of energy during a deformation, which subsequently ends in fracture. The aim of polymer modification is usually to develop a material with a high toughness and a large plastic elongation at break, whilst retaining the desirable properties such as

          DOI: 10.1002/masy.200451005

stiffness and strength. This is possible only by modifying the polymer morphology in such a way as to promote a large number of local energy-absorbing plastic deformation processes on a micro- and sub-micrometer scale.

It has been well known for many years that thermoplastics can be toughened by adding 5-25% of a suitable rubber.[1] This process of rubber toughening is of major importance to the plastics industry. It has proved so effective that the technology has been extended to almost all of the commercial glassy thermoplastics, including polystyrene (PS), poly(styrene-acrylonitrile) copolymer (SAN), polymethyl-methacrylate (PMMA) and polyvinylchloride (PVC). It has also been applied to several thermosetting resins and to semicrystalline polymers, notably polypropylene (PP) and polyamides (PA).

In rubber-toughened, high-impact polymers, the thermoplastic component forms the matrix in which the rubber phase is dispersed as particles. A general disadvantage of this modification is a pronounced decrease in strength and stiffness due to the rubber content. However, a good balance of these properties in combination with other properties (e.g. transparency, flame retardancy) and a good processability is demanded for many applications of polymers. Recently, some new techniques and micromechanical mechanisms were found as alternative toughening mechanisms. An overview of these mechanisms is given in the next section.

Block copolymers allow the control of structure on the nanometer scale and enable the combination of good mechanical properties with other properties such as transparency, recyclability etc. Due to the connectivity of the component chains and their inherent chemical incompatibility, the dissimilar blocks prefer to segregate, which gives rise to a rich variety of ordered structures called microphase separated structures.[2-4] However, the spatial extent of phase segregation is limited by the connectivity of the chains and hence the periodicity of structures lies in the order of the gyration radius (R_g) of the copolymer molecules. Classically, different ordered nanostructures are achieved by changing the overall composition of the copolymer in the two-component block copolymers.

Block copolymers have long been used as compatibilizers in binary polymer blends. Due to their intrinsic transparency, these polymers find application in the preparation of attractive packaging films for foodstuffs. Styrene/butadiene block copolymers can be used as impact modifiers for general purpose polystyrene.

In addition to the variation of the overall composition of block copolymers, the morphology can be adjusted by changing the processing conditions (processing temperature, cooling rate etc. or solution casting using different solvents). Application of external fields such as shearing, electric fields etc. can align the microstructures in a preferential direction leading to anisotropy in their physical properties.[5-9] Another way of adjusting morphology is to assemble the polymer chains in the form of ABC triblocks, giving the copolymer molecules different topologies (linear, star-shaped or grafted chains), and interfacial structures (neat, tapered or statistical transition at the interface), i.e., a wide variety of architectures. For example, in styrene/butadiene systems, the morphology ranges from PB cylinders in a PS matrix, to alternating lamellae and co-continuous structures, and even styrene domains dispersed in a rubbery matrix in a narrow composition range ($\Phi_{PS} \sim 0.70$).[10]

The formation of alternating layers of polystyrene and polybutadiene instead of cylindrical morphology in the sample is associated with its highly asymmetric architecture. A clear deviation from the classical picture of block copolymer phase diagrams can be achieved by the modified molecular architecture.[9, 11-13] The orientation of the microstructures and the occurrence of lamellar structures have a pronounced influence on the mechanical and micromechanical deformation behavior of the block copolymers. The deformation of the lamellar block copolymers is accompanied by a well-defined yield point. After the expansion of the necking zone over the whole tensile specimen (drawing), an orientation hardening begins. In contrast to the whitening of the linear block copolymers having PS matrix during tensile testing, the lamellar star block copolymer shows no stress whitening. Moreover, below a critical thickness of PS lamellae of about 20 nm, a transition from the usual craze-like deformation of PS to a so-called *'thin-layer yielding mechanism'* is observed.[9] This ability of the thin PS lamellae to reveal large homogeneous plastic deformation is considered as a new route of toughening on a nanometer scale.

A better understanding of structure modification and the improvement of mechanical properties is a task of particular scientific and economic importance. Structure, morphology and mechanical properties are linked by the micromechanical processes of deformation and fracture.[14, 15] Recently, smaller structural details have become increasingly important for defined improvement of mechanical properties, with a shift from the details on the micrometer scale to those on the nanometer range. Improved knowledge of the micromechanical processes on these different length scales provides a very direct way of improving mechanical properties.

Micromechanical processes are usually highly localized. Recently, several different techniques have been applied to study micromechanical properties. Besides spectroscopic and scattering techniques, the techniques of electron and atomic force microscopy are particularly useful for direct determination of micromechanical properties of polymers (an overview of these techniques in ref. [1417]). Using special micro-tensile devices for the electron and atomic force microscopes, in situ deformation tests can be performed. Since these techniques also allow the study of morphological details, they enable us to investigate structure-property correlations in a very direct way.

In this work, the toughness enhancement of rubber-modified or particle-modified semicrystalline polymers is discussed in dependence on the plastic yielding processes of the lamellar, semicrystalline matrix parts at and between the modifier particles. These lamellar yielding processes are compared with similar processes in lamellae-forming styrene/butadiene star block copolymers.

Concepts of Toughness Enhancement

There are several different mechanisms by which the toughness of polymers can be enhanced at room temperature or at lower temperatures. Some of them are widely applied in the plastics industry, others are theoretical possibilities.

i. Rubber Particle Toughening (General Modifier Particle Toughening)

Rubber toughened thermoplastics were first manufactured in the late 1940s and have been studied very extensively. Fracture toughness can be increased by up to one order of magnitude by adding a small amount (usually 5-25%) of a suitable rubber or elastomer to the thermoplastics. This effect was initially utilized in PS and SAN by grafting to butadiene rubber, yielding high-impact polystyrene (HIPS) and acrylonitrile-butadiene-styrene copolymer (ABS), respectively. The addition of rubber particles promotes energy absorption through the initiation of local yielding. Fibrillation in both the crazes and the rubber membranes of the 'salami' particles is typical of HIPS; the formation of matrix crazes in combination with dilatational shear bands, where the voids are confined to the rubber phase, is typical of many grades of ABS. Craze-like dilatational shear bands are formed in toughened grades of PVC, PMMA, PP and epoxy resin.[1, 14, 18] Elastomeric particles can be distributed in the matrix in the form of homogenous particles etc. In systems with 'salami' or 'core-shell' particles, hard polymer sub-inclusions occasionally can be deformed by cold drawing ('core flattening') as an additional mechanism of energy absorption.[19]

ii. Particle Filled Polymers (Composites)

Inorganic filler particles with a small interfacial strength yield to debonding and cavitation during loading. The microvoids thus formed act as stress concentrators (like elastomer particles), and can therefore initiate local yielding processes and increase toughness. Preconditions are the particles, which are small enough (including nanoparticles) with a narrow size distribution, with a good separation and optimum distance between the particles in the matrix.[14, 20] Particle filled PP and PE systems are examples of this mechanism.

iii. Rubber Network Yielding

It is often assumed that rubber toughening is synonymous with the addition of rubber particles. However, there is an alternative and also very effective approach to rubber toughening of amorphous polymers, using rubber networks. This method involves small particles of a thermoplastic, e.g., PVC, being embedded in the rubber network to form a honey-comb structure, with very thin layers of rubber separating the thermoplastic particles. The rubber content is usually kept below 10 vol.-% and the blend might consist, for example, of a rubbery ethylene-vinylacetate copolymer (EVA) matrix encapsulating primary particles of PVC which are ca. 1 μm in diameter. When the rubbery network is stretched, the PVC particles yield and absorb energy. One critical parameter is the thickness of the rubber layers, which should not exceed ca. 30 nm.[14, 21, 22] A disadvantage of these systems and the reason why they are not used in the plastics industry is the sensitivity during processing with destruction of the network and its transformation into the usual rubber particle distribution.[14]

iv. Inclusion Yielding

A deformation mechanism similar to that observed in rubber network blends is 'inclusion yielding', in which stiff thermoplastic particles are distributed in a matrix with a slightly lower yield stress, e.g., SAN particles embedded in a polycarbonate (PC) matrix).[23] Under load, stresses transferred to rigid inclusions via the softer matrix can exceed the yield stress of the inclusions and, therefore, the particles are forced to deform plastically and absorb energy.

v. Thin Layer Yielding Mechanism

In lamellae-forming styrene/butadiene star block copolymers, a large homogeneous plastic deformation of PS lamellae was found instead of the usual crazing mechanism. This effect depends strongly on the thickness of the PS lamellae. Below a critical thickness (of about 20 nm) a transition appears from the usual craze-like deformation to the thin-layer yielding mechanism with a drastic increase of elongation at break of up to 200-300%.[9, 24] This new toughening mechanism based on an effect on nanometer scale yields high impact and transparent polymers. In the case of loading the material not in a parallel direction to the lamellae but in an oblique or perpendicular direction, a modification of the deformation mechanism appears with the formation of so-called chevron-patterns or fir-tree or fish-bone patterns, which are described in this work.

vi. Self-Reinforcement

It is well known that strong orientation of the macromolecules increases stiffness and strength of thermoplastic polymers, e.g., PP and PE.[25] Relaxation processes during cooling demand a fast solidification[26] or the preparation only of thin films or fibers. Using special compaction techniques, bulk material can be produced from these films or fibers with good toughness at a high level of stiffness and strength.[27, 28]

vii. Phase Transformation

Following a concept of toughening of ceramic materials due to the phase transformation mechanism, a theoretical estimation shows that a stress induced transformation from one crystalline phase into another one (e.g. transition from β- to α-crystals in iPP) could absorb energy.[29] However, as yet, there is no experimental confirmation of this mechanism in polymers.

Experimental

Materials and Sample Preparation

Three different kinds of elastomer-modified semicrystalline polymers were used:

- A binary blend was prepared by melt blending of a high-density polyethylene (HDPE) with an elastomer (a very low density PE, VLDPE) in a wt.- ratio of 80/20. The blend was mixed in a Brabender Plasticorder PL 2000 at 180 °C and 40 rpm for 8 min before being pressed to form plates.[30]

- HDPE and 10 wt.-% of a rubber (a stabilized blend of a SB rubber and natural rubber) was mixed in an extruder and prepared as thin films.

- Another rubber-modified sample used was a commercial reactor blend of PP containing 15 vol.-% ethylene-propylene block copolymer (EPR) from PCD Polymers, Linz.

To study the deformation processes in the semicrystalline lamellar matrix in more detail, β-nucleated isotactic polypropylene (Daplen BE 60, Borealis AG, Linz) was used. 1 mm thick sheets were produced using a hot press and subsequently a multistage crystallization procedure.

Lamellae forming asymmetric styrene/butadiene block copolymers (number average molecular weight ~ 100 000 g/mole and total styrene volume content of 0.74) were investigated. For the detailed analysis of lamellar systems, blends of lamellar star block copolymer and polystyrene homopolymer (number average molecular weight ca. 80 000 g/mole) having wide molecular weight distribution were studied (details on the materials used are in refs. [9, 11-13]). The investigated materials supplied by the BASF Aktiengesellschaft, Ludwigshafen were prepared by injection molding (mass temperature 250 °C and mould temperature 45 °C) and solution casting using toluene as solvent.

Techniques

Morphology: To study the morphology of the materials used, ultrathin sections were prepared by ultramicrotomy, or in selected cases by cryo-ultramicrotomy. The thin sections were examined by conventional 100 – 200 kV transmission electron microscopy (TEM). Selective staining using heavy metal compounds such as ruthenium tetroxide (RuO_4) or osmium tetroxide (OsO_4) was generally employed.

Micromechanical testing: To investigate deformation micromechanisms, several different methods were used in this programme:

- Deformation in a commercial tensile tester and studying in a scanning electron microscope (SEM). In order to obtain planar and flat surfaces, the deformed miniaturized tensile bars were epoxy embedded and microtomed down to the middle section of the bar using a Leica microtome with a metal blade. The surface layers (influence of skin morphology, scratches) were thus eliminated. Regions of interest corresponding to the different states of deformation were selected for SEM investigations. Subsequently, a modified permanganic etching technique according to Olley and Bassett[31] was applied to reveal the nanostructure of the deformed material. To avoid electrical charging, samples were sputter coated with a gold layer of 12 nm. Micrographs of the deformation structures were recorded using a Jeol JSM 6300 scanning electron microscope operated at 15 kV accelerating voltage.

- Preparation of 0.5 – 2 μm thick semi-thin sections from the bulk materials followed by deformation in a special tensile stage fitted to a 1000 kV high-voltage electron microscope (HVEM, in cooperation with Max Planck Institute of Microstructure Physics in Halle/S.) or in a tensile stage for the atomic force microscope (AFM).

-   Preparation of ultrathin sections for deformation in a straining device attached to a 200 kV TEM, which also has facilities for cooling and heating the specimen.

Details of the different electron microscopic techniques are given in ref. [16, 17]

Tensile Testing: Tensile testing according to ISO 527 was performed at room temperature (23 °C) using a universal tensile machine at a crosshead speed of 50 mm/min. Some of the specimens were tested using miniaturized dumbbell-shaped tensile specimens by means of a Minimat miniature materials testing device (Polymer Laboratories, UK) at a crosshead speed of 1 mm/min at room temperature.

Results and Discussion

Elastomer modified HDPE

The morphology of the blend HDPE/VLDPE (weight ratio 80/20) is shown in a 200 kV TEM micrograph of an ultrathin section in Figure 1. The dark image structures are caused by selectively stained areas of the specimen by ruthenium tetroxide. On the one hand, they show the amorphous areas between the bright crystalline lamellae with an additional contrast. On the other hand, the dark areas of a few 100 nm in diameter result from strongly stained segregated particles consisting of the elastomeric blend component. Within the accuracy limit of the measurement, the result matches the weight ratio of the 80/20 (corresponding to the volume ratio) quite well. It is typical that a few particularly thick crystalline lamellae of the matrix PE penetrate into the segregated elastomeric particles.

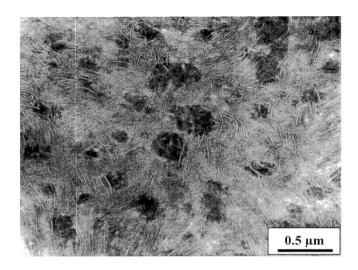

Figure 1.   TEM micrograph of a selectively stained ultrathin section of the blend HDPE/VLDPE
(weight ratio 80/20).

Using tapping mode in AFM, the specimen morphology in the surface region can be studied
without any chemical staining, as shown in Figure 2.   Due to their elastomeric properties, the
segregated particles appear bright and can be clearly distinguished from their surroundings
comprising crystalline HDPE lamellae.   Comparison of Figure 1 and Figure 2 reveals a segmental
appearance, like beads on a string, in the dark crystalline lamellae in AFM micrograph of
Figure 2. Figure 3 shows the image in the original state (1) and images of three steps of the tensile
test (2-4), corresponding to average local strains of 8, 16, and 56%.   To illustrate the local strain
distributions within the imaged (marked) area, four locations a, b, c, d are marked by arrows.
While the length of segment a (near the polar regions of elastomeric particles) does not change,
the length of segment b, which is located directly at an elastomeric particle (in the equatorial
stress concentration zone), increases significantly.   The corresponding local strain (at the three
deformation steps) was determined to be 16%, 30% and 146%, respectively.   This means that the
local strain at position b is much higher than the average strain. Segments c and d are arranged in
the matrix strand between elastomeric particles with a lower stress concentration. Therefore the

local strain is only slightly increased (8, 16, 69%, respectively).

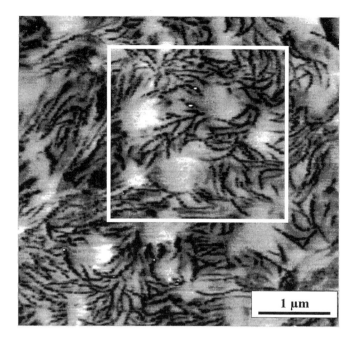

Figure 2.   Tapping mode AFM phase image of the HDPE/VLDPE blend with the area marked chosen for studying the micromechanical behavior of the blend.

The sequence of micrographs in Figure 3 represents a true in situ deformation test and illustrates the stress concentration effect at elastomeric particles. Besides the general effect of locally increased plasticity of the matrix strands between particles as the main energy absorbing and toughening effect, additional details of the yielding process are detectable. The area between the dotted lines in Figure 4a is deformed into the whole area visible in Figure 4c in the direction of tensile stress σ, corresponding to a mean strain of about 35%. The comparison of Figure 4a and 4c reveals the deformation behavior of the crystalline lamellae. The lamellae composed of crystalline blocks (like beaded strings) can be detected more easily in Figure 4b and 4d after additional image processing with a high pass filter.[30]

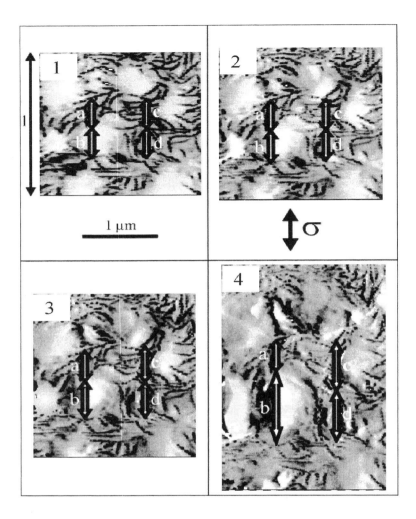

Figure 3.   Micromechanical processes in the marked area of Figure 2 shown by tapping mode
AFM phase micrographs showing the specimen areas before deformation (1) and in
the three stages of an in situ tensile test (2-4).

The typical behavior of lamellae is illustrated by selected lamellae segments marked by capitals at

their endpoints. Initially, the lamella segment between AB is nearly perpendicular to σ. During

the deformation, it rotates towards the stress direction by an angle of about 14° without changing the segment length. In contrast to that, the lamella segment CD and also the neighboring parallel lamella segment on the left are already oriented in the stress direction. Owing to the stretching of the lamellae the crystal blocks are separated from one another. The typical behavior of lamellae under stress results from the combination of both effects described. A lot of lamellar segments show rotation towards stress direction as well as additional stretch. This is illustrated by the lamella segment between E and F, which is located in the immediate vicinity of an elastomeric particle. This means that the interfaces between the blocks are the weakest points in the lamellar structure and show a preferred deformation with separation of the microblocks.

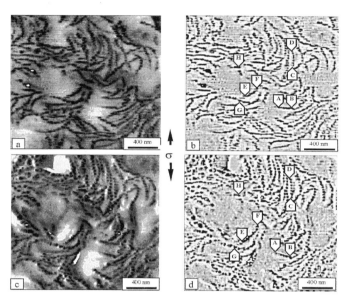

Figure 4. Tapping mode AFM images of the same area of the HDPE/VLDPE blend before deformation (a, b) and after deformation (c, d); pictures on the right (b, d) were produced by additional image processing of the original phase signals (a, c) with a high pass filter.

Pronounced twisting of lamellae is also detectable in the HDPE/rubber blend (see Figure 5). Around the rubber particles the crystalline HDPE lamellae are arranged in nearly perpendicular direction to the particle matrix interface and the extrusion direction MR – Figure 5. After deformation in the MR direction the longer lamellae are broken into shorter pieces, which are twisted into deformation direction, forming fir tree-like or chevron patterns – Figure 5b.

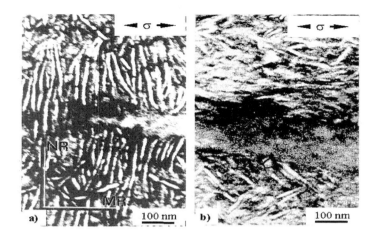

Figure 5.   Deformation structures in a HDPE/rubber blend: a) Extension ratio λ = 1: starting morphology with lamellae in perpendicular direction to extrusion MR; b) λ = 6: pieces of broken lamellae, twisted and oriented into deformation direction MR, forming fir-tree-like patterns; (NR – normal direction, MR – extrusion (machine) direction and deformation direction, stained thin sections in TEM).

Toughness enhancement of the blends is determined due to local stress concentration at and between the rubber particles and due to the yielding behavior of matrix strands between the particles. This yielding behavior is decisive for the toughening effect. The same mechanism of breaking, separation and twisting of lamellae can also be found during deformation of homo-PE[32] and β-nucleated iPP. These results correspond also with former findings about the deformation processes in HDPE at higher deformation temperatures below the melting point, revealing a greater number of defect layers or smaller crystalline microblocks inside lamellae oriented in parallel direction to the deformation direction.[33]

β-Modified Polypropylene

The special arrangement of the crystalline lamellae and interlamellar amorphous material in a parallel manner in β-iPP (stacks) can be interpreted as a lamellar nanostructure (10 to 20 nm) consisting of soft (amorphous) and hard (crystalline) components that are interconnected by tie molecules and entanglements. The characteristic deformation structures in β-iPP are shown in the electron micrographs in Figure 6.

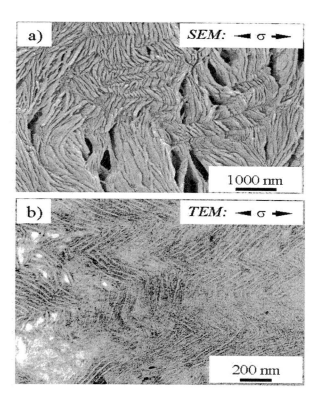

Figure 6.   Chevron structures of tilted and separated lamellae in a β-iPP at the early stage of deformation in direction of arrow: a) surface after permanganic etching in SEM and b) thin section after RuO₄ staining in TEM, deformation direction is shown.

A SEM micrograph of the deformed β-iPP (close to the necking region) is given in Figure 6a, where the straining direction is perpendicular to the direction of the lamellar stacks. In the initial stages of deformation ($\lambda \approx 1.2$), two main types of plastic processes can be distinguished. The formation of chevron-like morphology (zig-zag pattern) due to collective twisting of lamellar stacks and a lamellar separation process are observed simultaneously. Whereas the former process does not include a change in sample volume, the latter is accompanied by intensive microvoid formation and fibrillation within the amorphous portion that can be interpreted as a craze-like deformation mechanism (details in Figure 6b). From the measurement of average lamellar thickness it was concluded that in the initial deformation stage represented here, the crystalline lamellae (appearing as white strands in the TEM micrograph in Figure 6b) remain intact. The simultaneous processes of chevron formation and lamellar separation are controlled by the mobility of the interlamellar, amorphous portion of the material. Additional details about this mechanism can be studied using amorphous lamellar styrene/butadiene block copolymers, in which the thickness of PS and PB lamellae lies in the range of 10-20 nm (see below)

SBS Block Copolymers

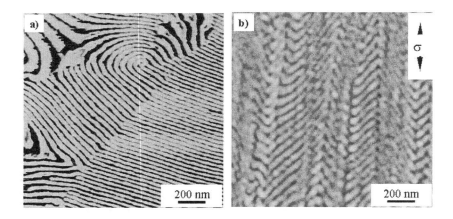

Figure 7.   AFM phase images showing the morphology of a solution-cast lamellar SBS triblock copolymer: a) without deformation and b) after deformation, perpendicular direction to lamellae arrangement; cryo-microtomed surface imaged in tapping mode.

It is known that the mechanical and micromechanical properties of block copolymer systems are strongly affected by the type, dimension and orientation of the microstructures. Additionally, the micromechanical response of these materials is influenced by the loading direction relative to the orientation of the microphase-separated structures.

If the material is loaded perpendicular or at an angle to the lamellar orientation direction, the lamellae are folded in a fish-bone-like arrangement (chevron morphology). The morphology of a lamellar SBS triblock copolymer film cast from solution is shown before and after deformation in AFM phase images, Figure 7. After deformation, the regions where the lamellae were initially perpendicular to the strain direction turn into so-called 'chevron morphology' or fish-bone morphology (zig-zag pattern).

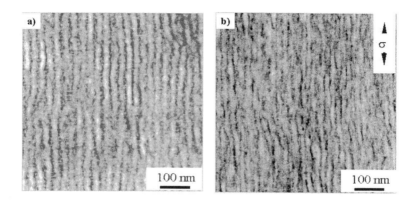

Figure 8.   Representative TEM images showing the morphology of an injection molded lamellar SBS triblock copolymer: a) without deformation and b) after deformation in parallel direction to lamellae arrangement; thin section selectively stained by OsO$_4$.

A different picture appears if the samples are loaded in such a way that the lamellae are parallel to the deformation direction. Figure 8 shows the morphology of an SBS star block copolymer before and after tensile deformation along the injection direction (or lamellar orientation direction). In the undeformed sample, the thickness of the PS lamellae and the lamellar spacing lies in the range of 20 nm and 42 nm, respectively, (Figure 8a). The tensile deformation parallel to the injection direction (i.e., the lamellar orientation direction) led to an extreme plastic drawing of both PS and PB lamellae, (Figure 8b). In the deformed sample the thickness of the PS lamellae and the lamellar spacing have been reduced to about half of their values before deformation. It is worth mentioning that the lamellae were stretched to a very high degree without any cavitation or microvoid formation. In contrast to the diblock copolymers, where the deformation localisation in

64

the craze-like zones is the principal deformation mechanism,[34, 35] no local deformation bands were observed.

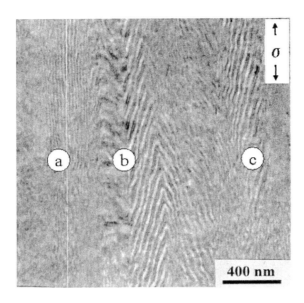

Figure 9.   TEM micrograph showing the deformation structures in a lamellar star block copolymer prepared by solution casting, strain direction is shown by an arrow; osmium tetroxide staining.

A summary of the deformation processes illustrated in Figure 7 and Figure 8 is collected in a TEM image of a solution-cast star block copolymer in Figure 9. If the deformation direction is perpendicular or oblique to the lamellar orientation (regions **b** and **c** in Figure 9), different mechanisms may act simultaneously or consecutively. Firstly, a shift of adjacent lamellae (gliding process) occurs. Thereafter, the lamellae break into smaller domains, rotating towards the loading direction. Finally, chevron-like or fir-tree like morphologies are formed. Formation of chevron morphology under mechanical loading was recently reported in oriented lamellar block copolymer samples subjected to tensile deformation perpendicular to the lamellar alignment [6]. It should be noted that the thickness of PS lamellae (about 20 nm) remains practically unchanged in the chevron folded structures after deformation. However, the lamellar long period is increased due to widening of the PB lamellae in the folded 'hinges'. Moreover, the thickness reduction of the PB layers during the parallel deformation was found to be more pronounced. These observations indicate that the rubbery phase reacts earlier towards the applied stress, which was further supported by Fourier transform infra red (FTIR) spectroscopy results.[36]

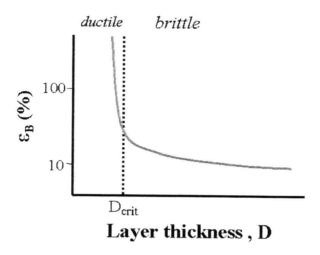

Figure 10.   Scheme showing the principal of thin layer yielding mechanism.

If the deformation direction is parallel to the lamellar orientation, a homogeneous plastic yielding of both PS and PB lamellae is detectable (region **a** in Figure 9). Apparently there is no change in the phase morphology besides a significant reduction of the layer thicknesses and the long period (compare different regions in Figure 9). The lamellae are strongly aligned towards the strain direction and the well-defined lamellar structure is even partly destroyed. The large plastic deformation of the glassy PS lamellae under tensile strain found in the star block copolymer studied is in line with the results reported earlier by Kawai et al., who observed a co-operative, drawing, shearing and kinking of the block copolymer microdomains.[37]

From the reduction of the lamellar spacing and the thickness of the PS lamellae in the undeformed and deformed samples, a local deformation of about 300 % can be estimated. This is the same order of magnitude as the craze fibrils stretching in the crazes of PS (craze fibril extension ration, $\lambda= 4$ [14, 38]). In other words, the yielding of the lamellae in the star block copolymer is analogous to the drawing of craze fibrils in the polystyrene homopolymer.

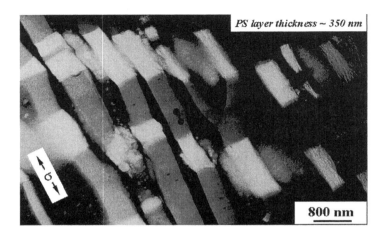

Figure 11. TEM image of a blend consisting of an SBS triblock copolymer and standard polystyrene (20/80 by weight) after deformation; the crazes are exclusively localized in the PS layers; deformation direction is shown.

The polymers designed to form the layered structures (e.g. alternating glassy and rubbery layers in the block copolymers) may undergo homogeneous plastic deformation even without forming any localized deformation zones. This behavior endows the polymers, which are otherwise brittle in the bulk state (e.g., bulk polystyrene), with a ductile property. This mechanism of homogeneous plastic deformation of PS lamellae (thin PS layers) together with adjacent PB lamellae can be described by a new deformation mechanism called '*thin layer yielding*',[9] which is schematically represented in Figure 10. This effect appears if the thickness of the PS layers lies below a critical thickness (D_{crit}). This critical thickness is comparable to the maximum craze fibril thickness in polystyrene homopolymer, i.e., in the range of 20 nm. The difference between the craze fibril yielding and the yielding of the PS lamellae lies in the fact that the craze fibrils stretch between microvoids while the PS lamellae undergo yielding between PB lamellae. Thus, in this case, the

PB lamellae act similar to microvoids in PS crazes and do not hinder the plastic deformation of the glassy polystyrene layers.

The large plastic deformation of the glassy lamellae at room temperature under tensile loading conditions is limited to lamella thickness below a critical value. The deformation mechanism changes from homogeneous drawing of the lamellae to the formation of local craze-like zones when the average thickness of the PS lamellae increases to about 30 nm.[24] The TEM image in Figure 11 illustrates, for example, in a blend of an SBS triblock copolymer and polystyrene homopolymer, how the thickness of the glassy layers affects the deformation micromechanism. Here, the PS layer thickness (average thickness ~ 350 nm) clearly exceeds the critical value, showing only fibrillated crazes inside. These results provide additional evidence for the '*thin layer yielding*' mechanism.

Comparison of Micromechanisms

Similar micromechanisms are observed in two entirely different classes of materials (PP and PE versus block copolymers), which result from their comparable morphologies (lamellar arrangement of the nanostructures). Depending on the loading direction relative to the lamellar alignment, different micromechanisms exist - as shown schematically in Figure 12. In the semicrystalline polymers, when loaded parallel to the lamellar orientation direction, an interlamellar gliding with plastic yielding of the amorphous layers between the lamellae and separation of the lamellae into smaller crystalline fragments with chain unfolding appear, leading to the formation of microfibrils at very high deformation. On the other hand, the mechanism of '*thin layer yielding*' dominates. In lamellar SBS block copolymers, interlamellar gliding is followed by plastic yielding of the butadiene and styrene lamellae corresponding to the "*thin layer yielding*" mechanism.

When loaded perpendicular to the lamellar orientation direction, similar deformation structures are formed both in the semicrystalline polymers (β-iPP and HDPE) and the lamellar SBS block copolymers, i.e., formation of chevron structures. In HDPE and also in β-iPP cavitation and fibrillation in the amorphous phase occur at larger deformation

The high plastic deformation of the amorphous region (including the formation of fibrillated crazes) in iPP leads to a high ductility of the sample. The special morphology of the β-form of iPP yields to a greater toughness than that of the α-form. The parallel array of crystalline lamellae enhances the local interlamellar shearing (if the lamellae are parallel to the loading direction) and lamellar separation and crazing of the amorphous phase (if the load acts perpendicular to the lamellar orientation direction). Since the amorphous region contains a number of defects (such as chain ends) it is more prone to cavitation, leading to the fibrillated crazes. The absence of microvoids in the soft phase of the SBS block copolymers is a most striking difference. The butadiene phase covalently bonded to the styrene chains (no molecular defects) allows cavitation only after the chain scission.

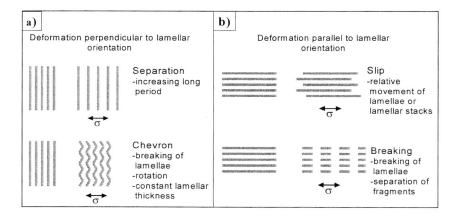

Figure 12. Scheme of micromechanical processes observed in lamellar systems investigated: a) lamellae perpendicular to the strain direction and b) lamellae parallel to strain direction.

Conclusions

Rubber toughening of amorphous and semicrystalline polymers is of major importance to the plastics industry. The basic effects are stress concentration at the particles and initiation of plastic yielding at and between the particles. In the past, enhancement of the toughening effect was achieved via optimization of the rubber particle size or the inter-particle distance. Up to now, an improvement of the yielding behavior of the matrix itself has not been discussed. Comparison of the micromechanical behavior of semicrystalline polymers (HDPE, iPP) with that of lamellae-forming amorphous SBS block copolymers revealed several similarities. These can be used to improve yielding ability of the semicrystalline matrix strands between the modifier particles. Aspects of particular interest in this sense are:

- The soft interlamellar phase is decisive for the process of interlamellar yielding (if deformation direction is parallel to the lamellae orientation) or twisting, reorganization and chevron formation of the crystalline lamellae (if deformation direction is perpendicular to the lamellar arrangement).

- A controlled cavitation inside the amorphous interlamellar phase can improve the lamellar orientation processes.

- Intense plastic yielding of the crystalline lamellae itself is assisted by breaking of lamellae into shorter crystalline blocks and transformation into microfibrils.

The micromechanical behavior of lamellae-forming heterophase polymers (based on semicrystalline polymers and amorphous block copolymers) has been comparatively studied. It is concluded that the deformation mechanisms of a nanostructured heterophase polymeric system may be dictated by the nature and arrangement of these structures. The basic mechanisms show two stages. The initial stage is characterised by a plastic deformation of the soft phase with a reorganisation of the hard (glassy or crystalline) lamellae. The second stage is determined by the alignment of hard phase towards the deformation direction and plastic yielding. In this way, the knowledge gained using a set of polymers may be used to understand the deformation mechanisms of another set of polymers with comparable phase morphology.

Additionally, the homogeneous plastic flow of glassy phase (i.e. *"thin layer yielding"*) in the lamellar block copolymer has been studied in detail. The significance of this mechanism from a practical point of view may be found in the fact that it can be used as an alternative toughening mechanism in other brittle polymers.

Acknowledgements

The authors thank Dr. S. Rudolf and Dr. Giesemann, Institute of Macromolecular Chemistry at the Martin Luther University Halle-Wittenberg, for providing the PE blends; Borealis AG, Linz for supplying PP samples and BASF Aktiengesellschaft for the block copolymers. They also thank Prof. U. Gösele, Max-Planck-Institute of Microstructure Physics in Halle/S, for access to the 1000 kV high-voltage electron microscope. We are indebted to the Deutsche Forschungsgemeinschaft (DFG) for financial support. SH and RA acknowledge the research grants from the Max-Buchner-Forschungsstiftung.

[1]   C. B. Bucknall, *"Toughened Plastics"*, Applied Science Publishers, London 1977.
[2]   S. Bates, G. H. Fredrickson, *Phys. Today* **1999**, *2*, 32.
[3]   I. W. Hamley, *"The Physics of Block copolymers"*, Oxford Science Publications, Oxford 1998.
[4]   H. Hasegawa, T. Hashimoto, in *"Comprehensive Polymer Science, Suppl. 2"*, S. L. Aggarwal, S. Russo Eds., p. 497, Pergamon, London 1996.
[5]   A. Keller, J. A. Odell, in: *"Processing, Structure and Properties of Block Copolymers"*, M. J. Folkes, Ed., p 29, Elsevier Applied Science Publishers, London 1985.
[6]   Y. Cohen, R. J. Albalak, G. J. Dair, M. S. Capel, E. L. Thomas, *Macromolecules* **2000**, *33*, 6502.
[7]   A. Böker, H. Elbs, H. Hänsel, A. Knoll, S. Ludwigs, H. Zettl, V. Urban, V. Abetz, A. H. E. Müller, G. Krausch, *Phys. Rev. Lett.* **2002**, *89*, 135502.
[8]   C. C. Honeker, E. L. Thomas, R. J. Albalak, D. A. Hadjuk, S. M. Gruner, M. C. Capel, *Macromolecules* **2000**, *39*, 9395.
[9]   G. H. Michler, R. Adhikari, W. Lebek, S. Goerlitz, R. Weidisch, K. Knoll, *J. Appl. Polym. Sci.* **2002**, *85*, 683.
[10]   R. Adhikari et al. in this volume
[11]   R. Adhikari, G. H. Michler, T. A. Huy, E. Ivankova, R. Godehardt, W. Lebek, K. Knoll, *Macromol.Chem.Phys.* **2003**, *204*, 488.
[12]   K. Knoll, N. Nießner, *Macromol. Symp.* **1998**, *132*, 231.
[13]   R. Adhikari, R. Godehardt, W. Lebek, R. Weidisch, G. H. Michler, K. Knoll, (**2001**), *J. Macromol. Sci.: Polym. Phys.* **2001**, *40*, 833.
[14]   G. H. Michler, *"Kunststoff-Mikromechanik: Morphologie, Deformations- und Bruchmechanismen"*, Carl Hanser Verlag, München 1992.
[15]   G. H. Michler, *J. Macromol. Sci.: Polym. Phys.* **1999**, *38*, 787.
[16]   G. H. Michler, *Trends Polym. Sci.* **1995**, *3*, 124.
[17]   G. H. Michler, *J. Macromol. Sci.: Polym. Phys.* **2001**, 40, 277.
[18]   C. B. Bucknall, in: *"Polymer Blends: Vol. 2 – Performance"*, D. R. Paul, C. B. Bucknall Eds. P. 83, John Wiley & Sons, New York 2000.
[19]   G. H. Michler, C. B. Bucknall, *Plastics, Rubber and Composites* **2001**, 30, 110.
[20]   G.-M. Kim, G. H. Michler, *Polymer* **1998**, *39*, 5699.
[21]   G. H. Michler, K. Gruber, *Plaste & Kautschuck* **1976**, *23*, 496.

[22] G. H. Michler, *Polymer* **1986**, *27*, 323.

[23] J. Kolarik, F. Lednicky, G. Locatt, L. Fambre, *Polym Eng. Sci.* **1997**, *37*, 128.

[24] E. Ivankova, R. Adhikari, G. H. Michler, R. Weidisch, K. Knoll, *J. Polym. Sci: Polym. Phys.* **2003**, *41*, 1157.

[25] I. M. Ward (Ed): *"Developments in Oriented Polymers"*, Elsevier Applied Science Publishers, London 1987.

[26] G. W. Ehrenstein, Cl. Martin, *Kunststoffe* **1985**, *75*, 105.

[27] N. D. Jordan, D. C. Bassett, R. H. Olley, P. J. Hine, I. M. Ward, *Polymer* 2003, 44, 1133.

[28] R. Bjecovic, *Ph. D. Thesis*, Universität Gesamthochschule Kassel, 2003.

[29] J. Karger-Kocsis, *Polymer Eng. Sci.* **1996**, *36*, 203.

[30] G. H. Michler, R. Godehardt, *Cryst. Res. Technol.* **2000**, *35*, 863.

[31] R. Olley, D. C. Bassett, P. J. Hine, I. M. Ward, *J. Mat. Sci.* **1993**, *28*, 1107.

[32] G. H. Michler, *Phys. Stat. Soc. (a).* **1995**, *150*, 185

[33] G. H Michler, *Colloid Polym. Sci.* **1992**, *270*, 627.

[34] C. E. Schwier, A. S. Argon, R. E. Cohen, *Polymer* **1985**, *26*, 1985.

[35] R. Weidisch, G. H. Michler, in: *"Block Copolymers"*, F. J. Balta´Calleja, Z. Roslaniec Eds., p. 215, Marcel Dekker, New York 2000.

[36] T. A. Huy, R. Adhikari, G. H. Michler, *Polymer* **2003**, *44*, 1247.

[37] M. Fujimora, T. Hashimoto, H. Kawai, *Rubber Chem. Technol* **1998**, *51*, 215.

[38] E. J. Kramer, In *"Advances in Polymer Science: 52/53, Crazing in Polymers-Vol I"*, H. H. Kausch Ed., p. 5, Springer Verlag, Berlin, Heidelberg 1983.

Nanostructures and Functionalities in Polymer Thin Films

Manfred Stamm,* Sergiy Minko,* Igor Tokarev, Amir Fahmi, Denys Usov

Institut für Polymerforschung Dresden, Hohe Straße 6, 01069 Dresden, Germany

E-mail: stamm@ipfdd.de; minko@ipfdd.de

Summary: Several examples of self-organization in thin polymer films are considered for the fabrication of nanopatterned surfaces and nanoscopic objects. Mixed polymer brushes of statistical distribution at the surface, composed of two immiscible polymers covalently bonded to the substrate phase, segregate in the sub-micrometer scale. Interplay between lateral and perpendicular (sandwich-like) segregation effects switching behavior of the thin films upon exposure to different environments. The switching between brush morphologies is used for the fabrication of adaptive/responsive surfaces. Fabrication of nano-domains based on microphase segregation in block-copolymer systems is used for structures with lateral dimensions as small as 5-50 nm. The ordered copolymer structures are applied as templates for the fabrication of membranes, nanofibers and nanoparticles.

Keywords: block-copolymer; morphology; nanolayers; nanotechnology; phase-separation

Introduction

Patterned thin films and surfaces at nanometer scale are of interest for several possible applications, where also specific functionalities are desired. There are several different approaches by which to fabricate thin polymer films with nanostructures. We discuss here three different methods that can be used to produce nanopatterns and nanoparticles utilizing self-assembly in polymer systems.

Firstly, a quite versatile technique based on self-organization of polymers is the use of phase segregation phenomena. One example comprises using mixed polymer brushes of statistical distribution at the surface, which are covalently bonded to the substrate and consist of different chains with different functionality.[1] Grafting restricts the possibilities of lateral segregation, and phase separation structures are in the sub-micrometer scale. Interesting switching functionalities

DOI: 10.1002/masy.200451006

can be achieved with these mixed brushes. This is a result of a combination of lateral and perpendicular phase segregation of components.[2] Wetting experiments reveal that these surfaces can adopt a hydrophobic or hydrophilic behavior, where one or the other functionality is achieved after dipping in different selective solvents for the components.[3] Utilizing structures of different length scales in addition to hydrophilic and hydrophobic switching behavior, a so-called ultra-hydrophobic surface can also be generated.[4]

Secondly, fabrication of nano-domains based on microphase segregation in block-copolymer systems is useful for structures with lateral dimensions as small as 5-20 nm. Typical structural size depends on the dimension of microphase-separated structures. The structures in diblock copolymers are, for instance, of lamella, cylindrical or spherical type, depending on the composition of components.[5] One can use the ordered copolymer structure as a template for further processing. In this case it is necessary to remove one of the components, which leaves, for instance, cylindrical holes in the polymer film. Another possibility comprises the use of an additional third, low molecular weight component, which is selectively incorporated in one of the phases, and which can be removed by washing.[6] By appropriate selection of the third component it is possible to obtain an orientation of the cylindrical domains in a thin film, perpendicular to the surface, and to fabricate regularly patterned, chemically heterogeneous surfaces.[7] The hole diameter can be of nanometer size and the morphology is ordered laterally on a relatively large scale. Holes can be filled with metals and, by complete dissolution of the polymer matrix, thin metallic nanorods on the surface can be obtained.

Thirdly, self-assembly of diblock copolymers can lead to microphase separation. This can be extended to macroscopic scale upon large external fields to achieve anisotropic properties. To produce anisotropic block-copolymer structures they have to be macroscopically aligned. Alignment of block copolymer structures under large stress fields can lead to considerable orientation of the morphology over a large scale in the range of several μm or even up to millimeters. Other external fields, such as oscillatory flow and electrical field, have been used to induce alignment[8-11]. Application of shear field provides an efficient and versatile tool to achieve macroscopic alignment, as demonstrated by materials subjected to extrusion,[12] oscillatory and steady shear,[13,14] extensional flow[15] and roll casting.[16] The aligned block-copolymer complex with the third low molecular weight component can be used for the fabrication of relatively long polymer nanofibers when the third component is incorporated into the matrix-forming phase and then the third component is selectively dissolved.

Experimental

Fabrication of mixed brushes. Clean Si-wafers were treated with (3-glycidoxypropyl)-trimethoxysilane (GPS, from ABCR) from a 1% solution in anhydrous toluene for 14 h in a dry atmosphere. They were then washed twice with anhydrous toluene in a dry atmosphere and thrice times for 5 min in an ultrasonic bath with methanol. In the next step diethylamine was deposited on the surface of the Si-wafers from a 1.5% solution in ethanol for 2 h. The resulting samples were rinsed five times with ethanol. The acid chloride derivative of 4,4'-azobis(4-cyanopentanoic acid) (ABCPA, from Fluka) was prepared by adding a slurry of phosphorus pentachloride to a suspension of ABCPA in dry dichloromethane at 0 °C and mixing overnight at ambient temperature in dry atmosphere. The product (Cl-ABCPA), after crystallization from dry hexane at 0 °C, was washed and dried *in vacuo*. In the next step Cl-ABCPA was deposited on the surface of the Si-wafers from a 1% solution in dichloromethane, with ca atalytic amount of triethylamine at room temperature for 12 h. The resulting samples of Si-wafers with chemically attached initiating groups were rinsed with ethanol in an ultrasonic bath. Oxygen was removed from the solution of the monomer styrene or 2-vinylpyridine (5-6 mol/l) and 4,4'-azobis(isobutyronitrile) (AIBN, from Fluka) (5-9×10^{-4} mol/l) in THF using five freeze-pump-thaw-cycles. The samples of the Si-wafers with the chemically attached initiator were placed in a monomer solution under argon atmosphere in a glass flask. The flasks were immersed in a water bath (60±0.1^{0}C) for 12 h. In the next step the same procedure was used to graft the second polymer using the Si-wafers with the first grafted polymer. The ungrafted polymers were removed by a cold Soxhlet extraction after each polymerization step for 1 and 8 hours, respectively, using THF. Every step of the modification of the Si-wafers was controlled by ellipsometric measurement of the layer thickness.

Fabrication of nanomembranes. Polystyrene(M$_w$=35500)-*block*-poly(4-vinylpyridine)(M$_w$=3680) (PS-*b*-P4VP, M$_w$/M$_n$=1.06, Polymer Source, Inc.) and 2-(4'-hydroxybenzeneazo)benzoic acid (HABA, ≥99.5% pure, Fluka) were used in an equimolar ratio between P4VP-block and HABA and dissolved in 1,4-dioxane (Acros Organics). The solution was kept overnight to complete hydrogen bonding and filtered directly before use. Films of the PS-b-P4VP + HABA assembly were deposited by dip-coating on cleaned silicon wafers. Solution concentrations of 1–3 % w/v resulted in film thickness of 20-90 nm. Finally, HABA was washed out by rinsing in methanol (Acros Organics), yielding nanoporous films. The films were characterized by ellipsometry (the Multiscope Optrel, Germany), Atomic Force Microscopy (AFM, the Dimension 3100, Digital Instruments, Inc.) in tapping mode, and UV-vis spectroscopy (the Cary 100 Scan UV-Vis spectrophotometer, Varian, Inc.).

Fabrication of nanofibers. Poly(styrene-b-4-vinyl pyridine) PS-b-P4VP had a (Polymer Source, Inc,), polydispersity index of M_w/M_n=1.07. The molecular weights of PS and P4VP were Mw=21400 g/mol and 20700 g/mol, respectively. 3-n-pentadecyl phenol (PDP) was purchased from Aldrich (purity 90-95%). It was distilled under reduced pressure (10^{-3} bar at 185 °C) and then recrystallised twice with cyclohexane to obtain a higher degree of purity. PS-b-P4VP was dissolved in analytical grade chloroform. A stoichiometric amount of PDP was added to the solution (one alkyl phenol for each pyridine group in the P4VP block). The solvent was allowed to evaporate slowly. The PS-b-P4VP (PDP) film was then dried at 60°C under vacuum for 24 h. The experiment then was carried out with a self-constructed piston-type spinning device. The upper part of the device consisted of a drive train, a heated cylinder with piston, and a single-hole die. The diameter of the capillary hole was 0.3 mm and its length was 0.6 mm. A low-speed winder was fixed below the die to try to wind up the extruded fiber. For melting, the material was heated up to 120°C within the cylinder, and a volume flow rate of 0.156 cm³/min was adjusted by the drive train. The extruded fiber was quenched by air at room temperature and collected or wound up at 2-4 m/min.

Mixed Brushes

The "grafting from" approach via a two-step, surface-initiated radical polymerization[1] was used for the synthesis of mixed polymer brushes (MPB) of a high grafting density (0.05-0.15 nm^{-2}) and a relatively large grafted amount (10-120 mg/m^2). The lateral distribution of the grafting points of the two polymers is assumed to be random. The covalent bonding to a substrate restricts segregation of the polymer species to the scale of a mean chain end-to-end distance.[2] Segregation of two polymer species in a MPB occurs in lateral and perpendicular (sandwich-like) directions and can be controlled by external factors (solvent, temperature). Heating above the glass transition temperatures of the brush polymers causes segregation of the polymer with lower surface energy to the brush surface. A selective solvent swells a favourite polymer and pulls it to the brush top, while the non-favourite polymer collapses and occupies the bottom brush layer. A good non-selective solvent swells both polymers, hence both of them are present on the brush top. The perpendicular segregation in MPBs was studied with water contact angle measurements on polystyrene/poly(2-vinylpyridine) brushes (Figure 1). Toluene and ethanol are selective solvents for PS and P2VP, respectively, while chloroform is a good non-selective solvent. PS as the less polar polymer than P2VP (thus providing the lower surface energy) segregated to the brush surface upon annealing.

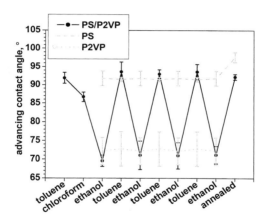

Figure 1. Reversible switching of a surface energetic state of a mixed PS/P2VP brush upon exposure to organic solvents of different selectivity and annealing for 24 h at 135°C. The behavior of monocomponent PS and P2VP brushes after the same treatment is shown as reference. The water advancing contact angle was determined with a sessile drop technique. The time of exposure of the brush to a solvent was 5 min; immediately after the exposure the brush was rapidly dried in nitrogen flux

Segregation in the lateral direction was studied with tapping mode AFM in the repulsive regime. A self-consistent field theory (SCF)[2] predicted for MPBs in solvents: (1) a ripple morphology composed of alternating lamellae of two brush polymers in a non-selective solvent, (2) a dimple morphology of round clusters of non-favorite polymer surrounded by a matrix of the favorite polymer in a selective solvent. Round clusters were found on PS/P2VP brushes after exposure to toluene and ethanol (Figure 2 a, c) and a worm-like structure (distorted lamellae) after exposure of these brushes to chloroform (Figure 2 b). The best phase contrast, originating from different viscoelastic properties of the polymers[17], was found for the brush exposed to chloroform. The sharp phase contrast after the non-selective solvent supports the theoretical assumption about the presence of both polymers on the brush top and is in good agreement with the contact angle values (Figure 1).

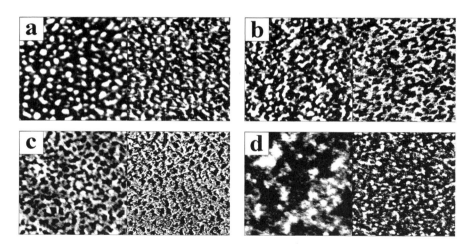

Figure 2. Repulsive tapping mode AFM of a PS/P2VP brush after exposure to the organic solvents (a,b,c) and annealing (d). Thickness 10 nm, 68 wt. % PS, Mw(PS)= 137000, Mw(P2VP)= 234000, Mw/Mn= 2,1 for both species. Left - topography, right – phase contrast.   Scale 1x1 µm. The images obtained after treatment of the brush with toluene (a), z scale 15 nm and 15°; chloroform (b), z scale 3 nm and 5°; ethanol (c), z scale 10 nm and 7°

Nanomembranes

Films of PS-b-P4VP + HABA assembly (Figure 3), prepared by dip-coating, are optically homogeneous and smooth, both macro- and microscopically. We found no contamination of the films by HABA crystals. Washing the films in methanol, which is a selective solvent for HABA, leads to disappearance of the characteristic absorption peak in the UV-vis spectra (382 nm), indicating the effective elimination of HABA from the films. The film thickness before and after washing in methanol was measured by AFM (a scratch test). No appreciable difference was found that gave an evidence of porous nature of the washed films.

AFM provides us with information on the surface topology of films of PS-b-P4VP. As prepared, the films are featureless, with rms roughness of 0.15 nm. Rinsing the films in methanol results in a drastic change of the surface topology. Independently, based on film thickness, AFM images (Figure 4, a) show close-packed, round pores, in quasi-hexagonal order, with a mean periodicity of about 21 nm. From the composition of the assembly, where P4VP + HABA is a minor block with a volume fraction of 25.5%, we can assume that this block forms cylinders in a PS matrix.[5]

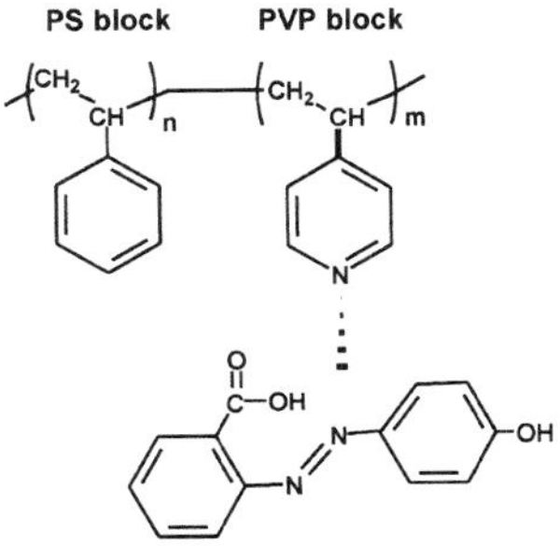

Figure 3.   PS-b-P4VP + HABA assembly

Elimination of HABA from microdomains of the minor block results then in the formation of cylindrical pores with P4VP brushes on the walls.[6] In AFM images the cylindrical pores appear oriented perpendicular to a film surface. They are arranged in a hexagonal lattice with a relatively narrow distribution of the distances between centers of neighbouring pores (the standard deviation is about 4 nm). There exists, however, no long-range lateral order as follows from an uniform ring of 2D FFT. We suppose that aggregation of HABA molecules attached to a P4VP block by hydrogen bonds plays a key role in the formation of the well developed, perpendicular aligned microdomain structure immediately after the deposition of the assembly.

Fast evaporation of solvent from the film surface during dip-coating results in "quenching" of film morphology in a partially ordered state. To impel further ordering we swelled films of the assembly in saturated vapors of 1,4-dioxane. A solvent in a swelled film serves as a plasticizer, inducing segment mobility.[18] Indeed, after vapor treatment we observed considerable improvement of the hexagonal order within relatively large domains (a few micrometers in size, see Figure 4, b). At swelling ratios of 2.5-3 (as monitored with *in-situ* ellipsometry) the best ordering was achieved. The mean distance between pore centers was 25±1 nm with a standard deviation of 2.2-2.5 nm.

We performed successful electrochemical growth of nickel dots into pores of the above described films (data are not presented here). This may be considered as a strong argument for cylindrical morphology of the films, where P4VP + HABA cylinders are aligned normal to the film surface.

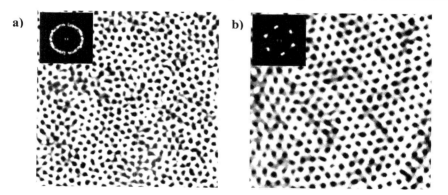

Figure 4.   AFM images 0(.5×0.5 μm²) and corresponding 2D FFT (as insets) of films of PS-b-P4VP + HABA assembly dip-coated from 1,4-dioxane: 40-nm thick, washed in methanol (a), 53-nm thick, swelled in vapors of 1,4-dioxane to a swelling ratio of 2.5, washed in methanol (b)

Nanofibers

The influence of the melt extrusion process of PS-b-P4VP diblock copolymer with addition of surfactant PDP on the morphology was investigated. Cylindrical nanostructures are formed by microphase separation in the melt. Large-scale orientation of PS cylinders within the P4VP/PDP matrix is studied by AFM, when only low flow rates are applied. After removal of PDP by ethanol, polymeric nanofibers are observed. The nanofibers are formed from a PS core with P4VP shell. These nanostructures can be produced in large quantities, because extrusion is a continuous process, and they can serve as templates for further modifications such as metallization. They may also be applied in nanocomposites to achieve enhanced mechanical properties. Self-assembly of block copolymers has been used to prepare individual polymeric nano-objects and nanostructures. Mäki-Ontto et al.[6] have demonstrated interesting ways of preparation of hairy tubes upon supramolecular organization in PS-PVP/PDP block copolymer complexes. The polymeric complex consisting of PS-b-P4VP, hydrogen bonded with PDP, shows a hexagonal morphology under equilibrium conditions, consisting of PS cylinders in the P4VP/PDP matrix. Melt extrusion and melt spinning of the material should increase the orientation of the PS-cylinders. A closer inspection of the filaments by AFM confirmed our assumption of oriented PS cylinders. PS cylinders had a thickness of about 90 nm in the P4VP/PDP matrix.

The final step in the production of nanofibers is the selective dissolution of the P4VP/PDP matrix. This process was already used by De Mole et al [19] to prepare PS nanofibers by using ethanol as solvent. However, with isotropic samples one usually may achieve nanofibers with a length of only about a few hundred nanometers.

Figure 5 shows PS nanofibers prepared by selective dissolution of the matrix and precipitation of the filaments on a substrate. The nanofibers exhibit lengths on the μm-scale, much larger than observed by previous authors, demonstrating the possibility to produce anisotropic nanofibers with considerable lengths. The cross section image in Figure 5 is used to determine the size of the nanofibers. PS nanofibers are smooth and macroscopically oriented, with considerable lengths. The width of the PS nanofibers is about 100 nm, slightly larger than observed in the P4VP/PDP matrix. This increase in diameter is due to the presence of a P4VP layer around the PS nanofibers.

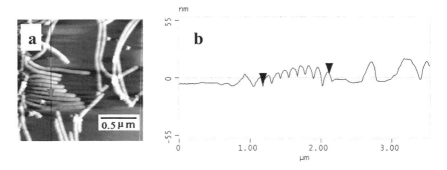

Figure 5.   AFM images of PS nanofibers coated with a P4VP layer prepared on a Si-substrate (a) and cross section (b)

Conclusion

We have shown that self-organization in polymer films can be tuned and adjusted to the fabrication of nanostructured materials for diverse applications. Mixed polymer brushes demonstrate an example of responsive materials with switched morphologies. Phase segregation in block copolymers is used for the fabrication of either regular arrays of nanochannels or for the preparation of polymeric nanofibers. Both of the materials can serve as templates for the fabrication of metallic nanorods and nanowires.

[1]   A. Sidorenko, S. Minko, K. Schenk-Meuser, H. Duschner, M. Stamm, *Langmuir*, **1999**, *15*, 8349.
[2]   S. Minko, M. Müller, D. Usov, A. Scholl, C. Froeck, M. Stamm, *Phys. Rev. Lett.* **2002**, *88*, 035502.
[3]   S. Minko, D. Usov, E. Goreshnik, M. Stamm, *Macromol. Rapid. Commun.*, **2001**, *22*, 206.
[4]   S. Minko, S.; M. Müller, M. Motornov, M. Nitschke, K. Grundke, M. Stamm, *J. Am. Chem. Soc.* **2003**, *125*, 3896.
[5]   I. W. Hamley *"The Physics of Block Copolymers"*, Oxford University Press, Inc., New York 1998, p. 26.
[6]   R. Mäki-Ontto, K. de Moel, W. de Odorico, J. Ruokolainen, M. Stamm, G. ten Brinke, O. Ikkala, *Advanced Materials* **2001**, *13*, 117.
[7]   I. Tokarev, S. Minko, M. Stamm, in *"Functional Nanostructured Materials Through Multiscale Assembly and Novel Patterning Techniques"*, S. C. Moss, Ed., MRS Proceedings, 2002, V. 728, 11-16.
[8]   Z. R. Chen, J. A. Kornfield, S. D. smith, J. T. Grothaus, M. M. Satkowski, *Science* **1997**, *277*, 1248.
[9]   R. H. Colby, *Curr. Opin. Coll. Int. Sci.* **1996**, *1*, 454.

[10] T. Thurn-Albrecht, J. Schotter, G. A. Kästle, N. Emley,T. Shibauchi, L. Krusin-Elbaum, K. Guarini,C. T. Black, M. T. Tuominen, T. P. Russell, *Science* **2000**, *290*, 2126.
[11] M. J. Folker, A. Keller, F. P. Scalisi, *Coll. Polym. Sci*, **1973**, *251*, 1.
[12] A. Keller, E. Pedemonte, F. M. Willmouth, *Coll. Polym. Sci*, **1970**, *238*, 25.
[13] G. Hadziioannou, A. Mathis, A. Skoulious, *Coll. Polym. Sci*, **1979**, *257*, 136.
[14] Y. Zhang, U. Wiesner, *J. Chem. Phys.***1995**, *103*, 4784.
[15] H. H. Lee, R. A. Register, D. A. Hajduk, S. M. Gruner *Polym. Eng. Sci.*, **1996**, *36*, 1414.
[16] R. J. Albalak, E. L. Thomas, *J. Polym. Sci. Polym. Phys.* **1993**, *31*, 3766.
[17] P. J. James, M. Antognozzi, J. Tamayo, T. J. McMaster, J. M. Newton, M. J. Miles *Langmuir* **2001**, *17*, 349.
[18] G. Krausch, R. Magerle, *Adv. Mater.* **2002**, *14*, 1579.
[19] K. de Moel, G. O. R. Alberda van Ekenstein, H. Nijland, E. Polushkin, G. ten Brinke, R. Maki-Ontto, O. Ikkala; *Chem. Mater.***2001**, *13*, 4580.

Macromol. Symp. **2004**, *214*, 85-96

Structure-Property Relationships in Rubber-Modified Styrenic Polymers

Walter Heckmann,* Graham Edmund McKee, Falko Ramsteiner

BASF Aktiengesellschaft, Polymer Research, D-67056 Ludwigshafen, Germany
E-mail: walter.heckmann@basf-ag.de

Summary: Styrenic polymers and copolymers are often impact modified with rubber particles. The efficiency of rubber toughening depends mainly on the size of the rubber particles and the degree of cross-linking. The deformation rate, the temperature, the orientation of the polymer molecules and the efficiency of rubber grafting also influence rubber toughening. It is thought that on impact, cavitation inside the rubber particles occurs which reduces the detrimental dilatational stress in the bulk polymer without forming cracks in the brittle matrix or at the rubber-matrix interface. Crazing and shearing are facilitated if the rubber particles can easily cavitate. This can be achieved by either avoiding too much cross-linking or by adding oil (silicone oil in the case of ABS) into the rubber particles, which acts as nuclei for void formation. An electron spectroscopic imaging method is described which allows visualizing the location of the oil. Already after cooling silicone oil modified ABS samples down to liquid nitrogen temperature rubber cavitation is observed. This cavitation is caused by the thermal stress developing due to the differences in thermal expansion coefficient between the rubber phase and the SAN-matrix and is facilitated by silicone oil. Voiding also leads to an increase of light scattering, which can be detected by an optical microscope using dark field illumination.

Keywords: ABS; electron microscopy; electron spectroscopic imaging; gloss; grafting; impact behavior; particle size distribution; rubber cross-linking; silicone oil

Introduction

Polystyrene (PS) and poly(styrene-co-acrylonitrile) (PSAN) are rigid polymers because they do not have a pronounced segmental mobility in the chain backbone below their glass transition temperature. Therefore they are brittle in impact testing. To improve this situation, the materials are often modified with rubber particles, leading to high-impact polystyrene (HIPS), acrylonitrile-butadiene-styrene (ABS) or acrylonitrile-styrene-acrylate (ASA). In these rubber-modified products, impact energy can be dissipated by crazing which is triggered in the stress field near the rubber particles. Crazing is favored by relatively large particles as demonstrated in the case of HIPS were crazes have formed perpendicular to the deformation direction (Figure 1a). In ABS, besides crazing, shear yielding is the predominant deformation mechanism if the rubber particles are small (usually of the order of 300 nm). These small particles favor a local plain stress situation in the region between the particles, with the consequence that a transition from crazing to shear deformation and macro-crazing can occur.

 DOI: 10.1002/masy.200451007

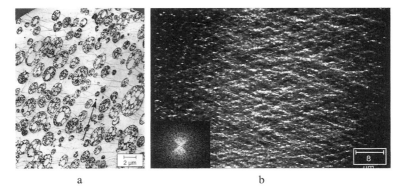

a   b

Figure 1.   Micrographs of a) crazes in HIPS (TEM) b) shear bands in rubber-modified PVC (optical micrograph and Fourier transform image).

In Figure 1b such a shear yielding behavior is shown for a deformed rubber-modified PVC sample. Crazing, macro-crazing and shear yielding are the most important deformation mechanisms for most polymers. Taking a closer look at a deformed HIPS sample shows that besides the crazes, voids are also visible within the rubber phase (Figure 2). The role and importance of void formation within the rubber particles is still not clear but it was recently studied by Bucknall et al. for ABS [1] and for rubber-modified PMMA by Béguelin et al.[2]. In both investigations it was assumed that void formation is an essential prerequisite for crazing.

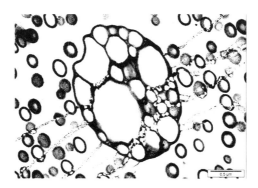

Figure 2.   TEM micrograph of a deformed HIPS sample.

The present paper will illustrate the importance of rubber particle size distribution, degree of rubber cross-linking, efficiency of rubber grafting and deformation temperature on the impact behavior of ABS. The importance of void formation and the influence of the addition of silicone oil on rubber cavitation will also be discussed.

The Efficiency of Rubber Toughening

A. Rubber particle size and deformation temperature

In Figure 3 TEM micrographs of two solution-polymerized ABS (LABS) samples with different particle size distributions are shown. LABS 321 has a 12% rubber content in the form of salami-like particles with a diameter up to 3 μm, whereas LABS 312 has a rubber content of 15% and a bimodal particle size distribution with smaller particles [3]. The

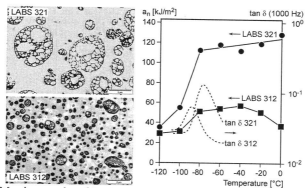

Figure 3.   TEM micrographs of LABS samples with different particle size distributions and corresponding impact strength a_n and damping (tan δ).

effectivity of rubber modification on impact energy is demonstrated in the impact strength, versus temperature curve (Figure 3). Obviously the sample with larger particles shows a higher impact strength which we think is due to a more effective craze-formation compared to the smaller particles. The smaller particles are too small to induce effective crazing and too large to induce shear yielding. What is also to be seen is that an increase in impact toughness is only observed when the deformation temperature is higher than the glass transition temperature of the rubbery phase which is between -70 and -90°C in the examples (mechanical damping below -60°C, ISO DIN 6721/3).

B. Rubber grafting

An important presupposition for good impact behavior is the necessity of sufficient rubber grafting. Normally the rubber particles must be surrounded by a closed grafted shell. In Figure 4a such a grafted shell is shown for HIPS where the grafted PS-shell around the rubber particles was made visible by dissolving the matrix polystyrene and the included PS with

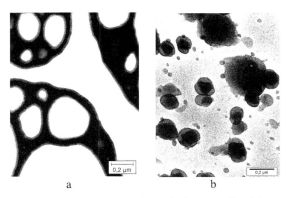

a b

Figure 4.   TEM micrographs of grafted shells around rubber particles in a) HIPS and b) ABS.

MEK/acetone directly on a TEM grid. The covalently bonded grafted PS molecules however cannot be extracted and so the grafted shell becomes visible. The grafted molecules give rise to a good adhesion between the PBu rubber particles and the PS matrix, and this is very important. The same is true for rubber particles in ABS prepared in emulsion, as shown in Figure 4b, where the rubber particles are grafted with SAN. In the literature it is discussed that a low degree of grafting and also a degree of grafting causes agglomeration of rubber particles during processing. This is undesired; it causes especially gloss and sometimes also mechanical problems [4]. Aoki [5] has discussed this problem by means of thermodynamic arguments. Due to insufficient grafting the incompatibility between PBu and PSAN leads to an increasing tendency to agglomerate. According to the model of Aoki too high a degree of grafting, where the grafted molecules are in a more extended chain conformation, will also cause agglomeration, because matrix chains are expelled from the grafted chains due to excluded volume effects. These two situations really cause agglomeration as was shown by experiment. Aoki found an optimal grafting level, which was specified at about 0.4 [5], where the grafting degree is defined as the ratio of the weight of the grafted SAN copolymer to that of the rubber particles. Entropic repulsion is assumed which suppresses rubber agglomeration. Hasegawa and coworkers [6] have pointed out that the graft density (number of grafted molecules per unit surface area of the particle) is the important parameter which influences rubber agglomeration. The optimum graft density was found to be about 0.08 nm^{-2} independent of particle content and particle size. This suggests that the optimum grafting density is determined essentially by the surface force between the particles. More recently Ahn and coworkers [7] found the optimum graft ratio at 0.6 where a good dispersion of the rubber particles was observed. When the ratio was 0.68 the particles tended to agglomerate but less severe compared to a lower graft ratio of e.g. 0.35.

C. Rubber cross-linking

In addition to the particle size and the degree of grafting the degree of cross-linking is also an important parameter, which influences toughness. In Figure 5 the torsion modulus of an emulsion ABS before and after annealing at 280 °C for 30 minutes is shown. Due to the annealing the glass transition temperature of the rubber phase has shifted by about 10 °C to higher values[8] and that means, that the degree of cross-linking is significantly higher in the annealed sample compared to the original sample. This additional cross-linking has a strong influence

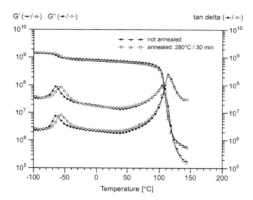

Figure 5.   Complex torsion modulus of ABS before and after annealing at 280°C for 30 min.

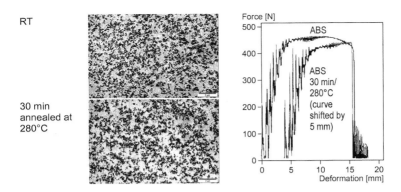

RT

30 min
annealed at
280°C

Figure 6.   TEM micrographs of undeformed ABS samples before and after annealing and corresponding force-deflection diagrams in impact strength test.

on the impact behavior, as can be seen from the force deflection diagram recorded during an impact test (Figure 6). To make it more clear the deformation curve of the annealed specimen was shifted along the deformation axis compared to the original sample by 5 mm. Thus the deformation at fracture is reduced by annealing or cross-linking from 15 to 10 mm. The loss

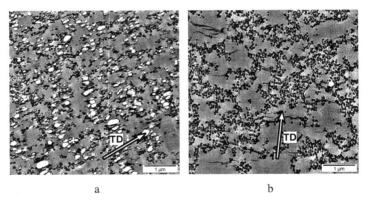

a   b

Figure 7.   TEM micrographs of deformed a) unannealed and b) annealed compression moulded ABS samples.

of toughness due to annealing appears to be caused by a transition in the deformation mechanism. Therefore the deformation process was studied by means of TEM on deformed samples. In Figure 7 a TEM micrograph of the original unanneald but deformed sample is compared to the corresponding annealed sample (Figure 7 b). In the non-annealed sample much void formation has occurred and so called macro-crazes have formed perpendicular to the tensile direction. These macro-crazes consist of voids and stretched fibrillar-like material between the voided particles oriented parallel to the tensile direction. We believe that void formation facilitates the shearing process between the cavitated particles in ABS. In the annealed case, however, cavitation is drastically reduced and only some conventional crazes have developed which bridge the particle free regions. Obviously due to cross-linking, the resistance to void formation has been increased and this must be the reason for the reduction in impact resistance to failure. This also demonstrates, that void formation in combination with shearing and stretching is an important event during the deformation process in ABS and crazing and shearing are facilitated by cavitation within the rubber particles.

D. Influence of silicone-oil on the impact behavior

Already in 1982 Morbitzer and coworkers[9] and recently also Bucknall[1] have shown that the addition of small amounts of silicone-oil can influence the mechanical properties of ABS. In Figure 8 the influence of a very small amount of silicone-oil on the impact resistance of ABS-

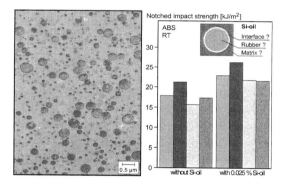

Figure 8.   TEM micrograph of an unmodified emulsion ABS sample and notched impact strength of normal ABS and ABS modified with silicone-oil.

samples is shown. Obviously with only 0.025% of silicone-oil a significant increase of the impact resistance can be observed. For a better understanding of this effect we tried to find out where the silicon oil is located. The oil could be located in the matrix phase (either segregated or dissolved on a molecular basis), in the rubber phase, again dissolved or as a separate phase or in the interface between the rubbery particles and the matrix. Dynamic mechanical measurements (Figure 9) reveal no influence of the silicone oil on the glass

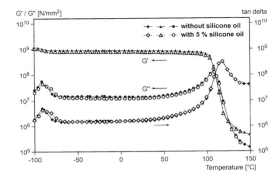

Figure 9.   Dynamic mechanical properties of ABS without and with 5% silicone oil.

transition temperature of the SAN matrix, here shown for a high silicone oil content of 5%. The glass transition temperature of the rubber phase, however, is broadened by the addition of oil to higher temperatures

To decide where the silicone oil is located we recorded a silicone elemental distribution map by means of electron spectroscopic imaging (ESI). This can be done by an analytical electron microscope where an energy filter is integrated in the column as realized in the LEO 912 Omega (Figure 10). The energy filter consists of magnetic prisms, which deflect electrons of

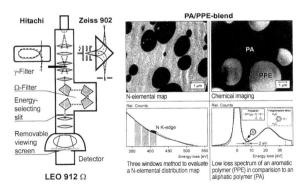

Figure 10. Detecting a nitrogen elemental distribution map by ESI.

different energy loss to different extents. In the back focal plane of the filter the electron energy loss spectrum is produced. With an energy selecting slit electrons of a special energy loss can be selected and used for imaging. Normally a total of three images with a 5 to 10 eV energy window are recorded. Two images in front of the edge are used to calculate the background, which is subtracted from the image under the edge resulting in the elemental distribution map. In Figure 10 this is shown for a N-distribution in a polyamide/ poly(phenylene ether) blend (PA/PPE)[10]. In the case of the ABS-samples we used the Si L2,3-edge for imaging because under the K-edge the signal was too weak. In Figure 11 the Si elemental distribution map of such a silicon oil-modified ABS sample is shown. For intensity reasons the silicon oil concentration was 0.43%. As it is clearly seen, the silicon oil has quantitatively concentrated inside the rubber particles and not in the interface or in the matrix. This is also clear by comparing the electron energy loss spectra on a rubber particle and in the PSAN matrix (Figure 11). Thus it is to be expected that any improvement of toughness by small amounts of silicone-oil is mainly caused by a promotion of deformation processes at the rubber particles. Bucknall[1] has shown that if an ABS sample is cooled down to liquid nitrogen temperatures and then warmed up to RT again, several voids inside the rubber particles of the ABS modified with silicone oil can be observed, whereas the rubber particles in an unmodified product are mainly unchanged. We did the same experiments and came to similar results (Figure 12). We find voids within the rubber particles of silicon oil modified

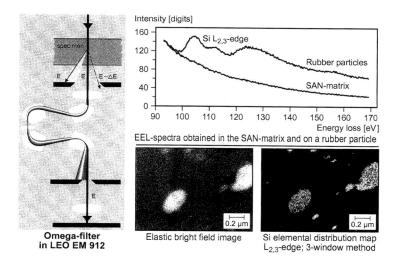

Figure 11. Silicone elemental distribution map of ABS modified with silicone oil obtained by ESI.

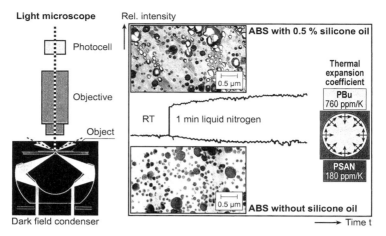

Figure 12. Intensity of light scattering before and after cooling silicone oil modified and unmodified ABS samples to liquid nitrogene temperature.

ABS after cooling down the sample to liquid nitrogen temperature followed by warming up to room temperature. There was no void formation in an unmodified sample. The voiding effect can also be followed by measuring the scattered light under a dark field condenser in an optical microscope [11, 12]. The scattered light intensity significantly increases in the silicone oil-modified sample and is nearly unchanged in the unmodified sample. Bucknall has shown, that due to the different thermal expansion coefficients between the rubber phase and the PSAN a strong stress is built up inside the particles and this stress can relax by voiding. This occurs more easily in the particles modified with silicone oil. This relaxation of the rubber particles by voiding is the reason why we observe a broadening of the glass transition temperature of the rubber phase to higher temperatures. Those relaxed rubber particles show a higher glass transition temperature compared to the particles still under stress (higher free volume)[1,9]. By cooling down to different temperatures and subsequently measuring the light scattering after warming to room temperature we have found, that void formation begins between -80 and -90°C, this is close to the glass transition temperature of the rubber phase (Figure 13).

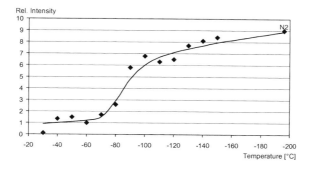

Figure 13. Relative intensity of scattered light of thin ABS films modified with silicone oil and cooled to different temperatures.

Void formation within the rubber particles will also happen during an impact test. Due to the silicone oil, cavitation within the rubber particles obviously will occur more easily compared to the unmodified ABS and so crazing and shearing are facilitated and a higher impact resistance is observed. In contrast to this, on thermal cross-linking the rubber particles become more resistant to cavitation, resulting in embrittlement.

Conclusion

It was shown, that the efficiency of rubber toughening in ABS depends on several parameters. Beside of the deformation rate, which was not discussed, the most important parameters are

- the particle size distribution
- rubber grafting
- degree of cross-linking
- deformation temperature
- ease of void formation

Especially void formation in rubber particles has a more important influence on the impact strength than was supposed in the past. However, it is still open to question, which of the following three mechanisms prevails, namely

- whether crazes are initiated after void formation
- whether the precursor of a craze in the matrix can only easily form a real craze, if the adjacent rubber particle cavitates to compensate the local volume strain or
- both deformation structures occur simultaneously

Void formation can be facilitated by adding an appropriate silicone oil whereby the silicone-oil is concentrated in the rubber particles as shown by ESI. Even small amounts lead to an improvement in toughness. Further important parameters are adequately grafted rubber particles of the correct particle size, which can internally cavitate but are strong enough to bridge deformation zones like crazes.

[1]   C. B. Bucknall, D. S. Ayre, D. J. Dijkstra, *Polymer* **2000**, *41*, 5937.
[2]   Ph. Béguelin, Ch. J. G. Plummer, H. H. Kausch, *"Polymer Blends and Alloys"*, G. O. Shonaike, G. P. Simon, Eds., Marcel Dekker, New York 1999, p. 549.
[3]   F. Ramsteiner, G. E. McKee, W. Heckmann, W. Fischer, M. Fischer, *Acta Polym.* **1997**, *48*, 553.
[4]   M. C. O. Chang, R. L. Nemeth, *J. Appl. Polym. Sci.* **1996**, *61*, 1003.
[5]   Y. Aoki, *Macromolecules* **1987**, *20*, 2208.
[6]   R. Hasegawa, Y. Aoki, M. Doi, *Macromolecules* **1996**, *29*, 6656.
[7]   K. H. Ahn, D. H. Ha, B. D. Lee, J. G. Doh, J. H. Choi, *Polym. Eng. Sci.* **2002**, *42*, 605.
[8]   L. E. Nielson, *J. Macromol. Sci. Rev.* **1969**, *69*, 77.
[9]   L. Morbitzer, G. Humme, K. H. Ott, K. Zabrocki, *Angew. Makromol. Chem.* **1982**, *108*, 123.
[10] W. Heckmann, W. Probst, F. Hofer, W. Grogger, *14th International Congress on Electron Microscopy*, Cancun 1998, Proceedings Vol. II, p. 867.
[11] W. Heckmann, *15th International Congress on Electron Microscopy*, Durban 2002, Proceedings Vol. I, p. 587.
[12] F. Ramsteiner, W. Heckmann, G. E. McKee, M. Breulmann, *Polymer* **2002**, *43*, 5995.

Microdeformation in Heterogeneous Polymers, Revealed by Electron Microscopy

Christopher J.G. Plummer,[1] *Philippe Béguelin,*[2] *Chrystelle Grein,*[2] *Rudolph Gensler,*[2] *Laure Dupuits,*[1] *Cedric Gaillard,*[3] *Pierre Stadelmann,*[3] *Hans-Henning Kausch,*[2] *Jan-Anders E. Månson*[1]

[1]Laboratoire de Technologie de Composites et Polymères (LTC), IMX-STI, Ecole Polytechnique Fédérale de Lausanne (EPFL), CH-1015, Switzerland

E-mail: christopher.plummer@epfl.ch

[2]Laboratoire de Polymères (LP), IMX-STI, Ecole Polytechnique Fédérale de Lausanne (EPFL), CH-1015, Switzerland

[3]Centre Interdisciplinaire de Microscopie Electronique (CIME), SB, Ecole Polytechnique Fédérale de Lausanne (EPFL), CH-1015, Switzerland

Summary: Recent investigations of the tensile fracture behaviour of representative glassy and semicrystalline impact-resistant polymers are reviewed, with emphasis on the microdeformation behaviour as revealed by electron microscopy of sections from bulk specimens, where the crack-tip damage zone has been embedded and/or stained under load, and on thin films deformed *in situ*. The insight that such techniques have provided into the toughening mechanisms is discussed.

Keywords: deformation; electron microscopy; fracture; high speed testing; polymers

Introduction

Polymers that require a high degree of fracture resistance are often combined with a second phase toughener. A classic example is high impact polystyrene (HIPS), a generic term for polystyrene (PS) that contains rubbery inclusions, whose role is to promote dissipative deformation and hence increase crack initiation and propagation resistance. Conventional impact tests provide a useful measure of fracture resistance at overall deformation rates equivalent to between roughly 1 and 4 m/s. Such tests are limited in scope, however, and there is growing interest in quasi-static tensile testing, which gives access to intrinsic fracture toughness parameters over a wide range of test

DOI: 10.1002/masy.200451008

speeds, including impact speeds, and hence facilitates the systematic investigation of rate effects [1]. A parameter commonly used is the critical stress intensity factor for crack initiation in plane strain mode I opening, K_{IC} (generally the most severe test of a polymer's fracture resistance). However, a working description of the impact response may also require an indication of whether crack propagation is stable or unstable, i.e. whether failure is ductile or brittle. Indeed, in many applications, the aim is not so much to prevent failure as to ensure that it occurs via stable crack propagation, as in the case of materials for airbag housing, where it is important to minimize damage to the airbag and/or the user from flying debris during deployment.

Although fracture mechanics can provide an objective description of crack resistance, it is also important to identify the underlying microdeformation mechanisms, particularly in heterogeneous impact polymers, where crack tip deformation is often strongly dependent on the modifier properties, content and geometry. In conventional polymer matrices, there are two main types of irreversible deformation: cavitational deformation in the form of crazes, and "shear deformation", which usually refers to isovolumetric, i.e. non-cavitational plastic, deformation. Crazing is a localized deformation mode, favoured by the hydrostatic stress concentrations around crack tips in thick specimens under load, whereas shear deformation must be cooperative throughout the specimen thickness in order for the constant volume criterion to be satisfied. The connectivity of the craze fibrils means that their failure strength may be reached at craze widths of only a few μm, so that if only one craze is present at the crack tip, say, little energy is dissipated during crack advance and failure is macroscopically brittle [2, 3]. In materials such as PS, in which crazing is the dominant microdeformation mechanism, the stated aim of toughening strategies is therefore often to promote multiple craze nucleation [4-6]. Cavitation or debonding of either rubber or mineral particles, and indeed crazing itself, may nevertheless relax triaxial constraints on the matrix material at the crack tip sufficiently to favour shear yielding, which may in turn provide an important contribution to energy dissipation [4-8]. Recent work has also suggested that in semicrystalline polymers such as polyamide 6 (PA6) and polyethylene (PE), modifier particles may orient the lamellae locally so as to reduce the effective yield stress [9-13].

In what follows, after a brief description of the experimental approach, the crack-tip deformation behaviour of two types of toughened polymer will be discussed and compared. The first is rubber-toughened glassy amorphous polymethylmethacrylate (PMMA), toughened with of a variety of tailored preformed composite particles. The second is semicrystalline isotactic polypropylene (iPP), whose T_g is below ambient temperature, but which also shows craze-like deformation and brittle impact behaviour, particularly at low molar masses. In addition to rubber toughening of iPP, which usually involves blending with an un-crosslinked modifier, efforts to improve intrinsic toughness by modification of its crystalline structure will be discussed.

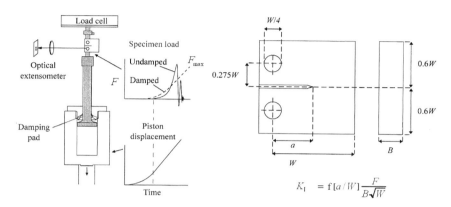

Figure 1.   Set-up for high-speed testing with the CT geometry (left hand side), and the method for calculating K_I from the data (a is the crack length, F is the force, B is the specimen thickness, W is the specimen width and f[a/W] is a geometrical factor related to the specimen geometry).

Experimental

Mechanical testing: Mode I fracture tests have been carried out on compact tension (CT) specimens using a Schenck Hydropuls POZ 1152 servo-hydraulic test apparatus (Figure 1). At test speeds, $v > 0.1$ ms^{-1}, a damping pad attenuates dynamic effects, and so maintains quasi-static conditions up to the highest v (about 10 ms^{-1}) [1, 14, 15]. A mode I stress intensity factor, K_I, may then be calculated as indicated in Figure 1. If the specimen response is linear elastic up to the maximum force, F_{max}, substitution of F_{max} for F in the expression for K_I gives K_{IC}. The ductile-brittle transition is defined as the transition from stable or partly stable crack propagation (that is, $F > 0$ at displacements beyond that corresponding to F_{max}), to unstable crack propagation. In the regimes of relatively ductile behaviour that accompany stable crack propagation, K_{IC} may be difficult to determine rigorously for practically attainable specimen thicknesses, owing to curvature of the force displacement curves [16]. The nominal critical stress intensity derived from the force maximum nevertheless provides a conservative measure of the crack initiation resistance, and remains useful for comparing different materials and the influence of materials characteristics and test parameters. Therefore, for the present purposes, no distinction is made between "valid" and "invalid" K_{IC} data.

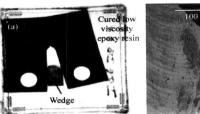
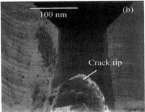

Figure 2.   Specimen preparation: (a) deformed specimen with a wedge in place, mounted in epoxy resin; (b) region around the crack tip trimmed for sectioning and SEM or TEM observation [17].

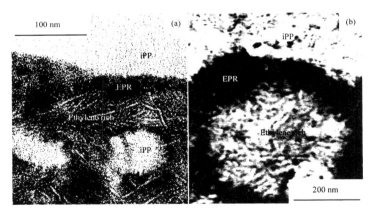

Figure 3.   Modifier morphology in iPP blended with EPR: (a) TEM (Philips EM430, 300 kV, bright field, RuO$_4$ stained) (b) LV-SEM (Philips XL30 SFEG, 3 kV, TLD, RuO$_4$ stained, contrast inverted to facilitate comparison with the TEM image).

Morphological characterisation: Because crazes and other types of cavitational deformation undergo relaxation and collapse on unloading, they may need to be embedded in a low viscosity epoxy or acetate resin under load and/or marked using a suitable stain prior to sectioning for transmission electron microscopy (TEM) [14, 17, 18]. This is conveniently achieved in notched specimens by inserting a wedge between the crack faces (Figure 2). Similar techniques may also be used for low-voltage, high-resolution scanning electron microscopy (LV-SEM) [19-21]. Indeed, LV-SEM [22] and atomic force microscopy (AFM) and related near-field techniques [23] are now challenging the dominance of TEM at the length scales of the order of 10 nm that characterise not only microdeformation (voids, fibrils) but also structural components of semicrystalline and impact toughened polymers (crystallite thicknesses, modifier particles). If a backscattered electron (BSE) detector or "through-the-lens" detector (TLD) is employed in conjunction with LV-SEM, for example, staining provides sufficient chemical contrast to permit identification of individual lamellae, as illustrated in Figure 3, which compares TEM and LV-SEM images of the modifier morphology in a rubber toughened iPP.

Thin film techniques: TEM of thin, electron transparent films, deformed *in situ* by straining a copper support, which maintains them under load in the microscope, has been used extensively to investigate crazing in amorphous polymers [24]. Although easily adapted for SEM and AFM, such techniques are limited by their incompatibility with deformation on length scales comparable to or greater than the film thickness, e.g. coarse cavitation in semicrystalline polymers [25, 26], and difficulties in obtaining microstructures representative of bulk polymers with *ad hoc* preparation methods such as solvent casting. It is therefore often of interest to prepare thin sections directly from bulk polymers by ultramicrotomy [8, 27, 28]. To deform thin films at impact speeds, use has been made here of a simple apparatus based on the same principle as for the high-speed testing of bulk specimens, i.e. quasi-instant acceleration of the specimen fixture to the required speed using a pre-accelerated falling weight [8].

Rubber-toughened PMMA

PMMA is an amorphous thermoplastic with good optical characteristics, but in which crack propagation during tensile failure of notched specimens is typically characterised by a single crack tip craze and hence relatively brittle behaviour. This is particularly true at high deformation rates, reflecting a trend towards crazing at high v and low T, whereas shear deformation dominates as T approaches T_g [29]. The challenge is to improve impact resistance in PMMA without sacrificing optical properties, so that it may replace polymers that are tougher, but more expensive, or inorganic glasses in applications such as car windows, where safety requires substantial energy absorption during impact. Emulsion polymerization offers considerable design freedom for modifier particle morphology and size, which can be fixed by crosslinking during synthesis, and hence maintained on dispersion in a PMMA matrix or matrix precursor [30-32]. Homogeneous dispersion and particle-matrix adhesion are assured by the presence of a graft PMMA outer shell. In commercial grades, the composition of the layers is adjusted to ensure transparency [30].

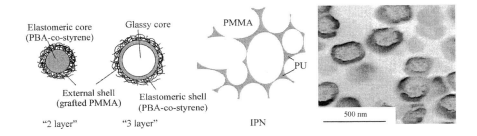

Figure 4. Modifier morphologies in rubber-toughened PMMA, along with a LV-SEM micrograph of a specimen that contains 47 wt.-% 3 layer particles (3 kV, TLD, RuO$_4$ stained, contrast inverted).

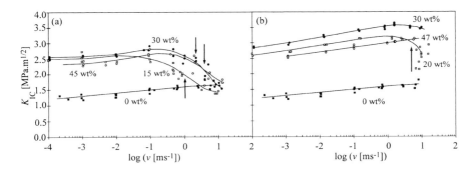

Figure 5. K_{IC} as a function of v in (a) 2 layer (b) 3 layer particle toughened PMMA for different particle contents: the PMMA matrix (0 wt% modifier) is brittle throughout whereas fully ductile behaviour is observed with 30 and 47 wt% 3 layer particles, and the remaining materials show a ductile-brittle transition indicated by the arrows [14]. (Matrix M_w = 130 kg/mol, optimized particle diameters (160 and 250 nm for the 2 and 3 layer particles respectively).

Typical modifier morphologies are shown schematically in Figure 4, along with a micrograph from a bulk specimen of toughened PMMA. Modifier particles in which the rubbery phase is restricted to a shell ("3 layer particles") give better stiffness than an equivalent loading of particles with rubbery cores ("2 layer particles"). Moreover, as illustrated in Figure 5a and b, which give room temperature results for K_{IC} as a function of v, 3 layer particles may also lead to significant improvements in K_{IC}, and a displacement of the ductile-brittle transition to higher v at fixed modifier content [33-35]. Thus, although 2 layer particles give little or no improvement over the neat matrix at the highest v, 3 layer particle contents of 30 wt% or more result in stable crack propagation over the whole range investigated [8, 14].

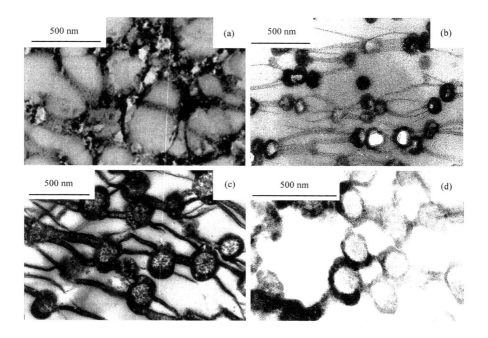

Figure 6.   TEM of crack-tip deformation in CT specimens of different types of toughened
PMMA deformed at about 1 mm/s at room temperature: (a) cavitation in the rubbery
phase of an IPN; (b) cavitation and crazing in a specimen containing 15 wt% 2 layer
particles; (c) cavitation and crazing in a specimen containing 30 wt% 3 layer
particles; (d) cavitation and crazing in a specimen containing 15 wt% 3 layer particles
(all specimens stained in RuO_4 prior to sectioning, tensile axis roughly vertical in
each case) [8, 14].

Microdeformation in bulk specimens: Direct observations of crack tip deformation are shown in
Figure 6 for different toughened PMMAs deformed at 1 mm/s, i.e. well within the regime of
stable crack propagation. Also shown is crack-tip deformation in a crosslinked PMMA-based
interpenetrating network (IPN) in which the modifier phase (polyurethane, PU) forms a
continuous network, shown schematically in Figure 4 [36]. In the particle toughened PMMAs,
stable crack propagation is accompanied by intense stress whitening, due to the formation of a
network of cavitated particles and crazes. Crazing does not occur in the IPN, however,
presumably owing to crosslinking, which hinders cavitation in the matrix [2]. The observation of
comparable values of K_{IC} in the IPN and PMMA/30 wt% 3 layer particles and fully stable crack

propagation [8] therefore points to the primordial role of shear deformation in energy dissipation at the crack tip. Cavitation of the IPN rubbery phase (about 6 wt% of the material), visible in Figure 6a, is expected to lead to similar constraint release and crack shielding to those resulting from the formation of a continuous network of cavities and crazes in the particle toughened PMMA [8, 14, 35]. For the PMMAs with relatively low particle contents, on the other hand, crazing and constraint release are less efficient, leading to more brittle behaviour. Moreover, whereas cavitation of the rubbery shells of 3 layer particles results in load-bearing craze-like fibrillar structures uniformly distributed among the particles (as reflected by distortion and/or fibrillation of the particle cores at high deformations), cavitation of the 2 layer particles is relatively non-uniform. The corresponding large voids in the cores of certain of the 2 layer particles (Figure 6b) have been suggested to constitute weak spots in the deformed material [28].

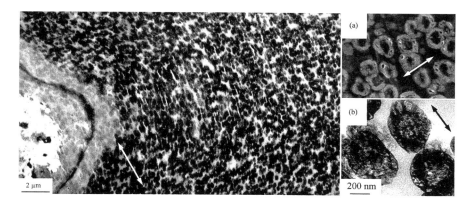

Figure 7.   Crack tip deformation in a chemically modified PMMA reinforced with 47 wt% 3 layer particles and deformed at 0.1 mm/s; the insets show details of particle deformation by (a) LV-SEM and (b) TEM (all specimens stained with RuO$_4$, tensile axis indicated by the arrows) [37].

At particle contents approaching 50 wt%, cavitation remains prevalent in the shells of the 3 layer particles, but crazing is limited, and extensive shear deformation may be inferred from the distortion of the modifier particles near the crack tip. Indeed, in this case, constraint release and the increased matrix ligament stresses are sufficient to reduce the overall effective yield stress, and hence the nominal K_{IC}, as the particle content increases from 30 wt% to 47 wt% (Figure 5).

Similar qualitative behaviour is seen in a more ductile chemically modified PMMA matrix with twice the entanglement density of PMMA and $T_g \approx 85$ °C, as compared with 100 °C for PMMA. Here, crazing is fully suppressed in favour of shear deformation at low v (Figure 7). Indeed, in spite of the lower T_g, K_{IC} is slightly greater than that of the conventional PMMA matrix at the same loading [37], suggesting crazing to have a negative effect in this limit.

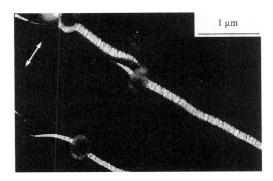

Figure 8.   TEM of a thin film containing 15 wt% 3 layer particles deformed at about 1 m/s [29].

Thin film microdeformation: Among the disadvantages of direct observation of microdeformation in bulk specimens is the need to stain to obtain adequate contrast, which obscures details of the morphology. Given the relatively small length scales involved in the microdeformation of PMMA toughened with 3 layer particles, *in situ* deformation of thin films is therefore a useful complement to bulk studies. Voided regions (either modifier particles or crazes) may easily be seen by TEM without staining, provided low dose conditions are used to avoid beam damage. It is also straightforward to carry out systematic observations over a wide range of conditions using thin films, since no additional specimen preparation is required. Figures 8 and 9 give examples of deformation in films of about 300 nm in thickness prepared by microtoming bulk specimens. Figure 8a shows a thin film containing 15 wt% 3 layer particles deformed at 1 m/s showing isolated crazes with relatively straight trajectories, corresponding to the onset of brittle behaviour in the bulk fracture specimens (the undeformed particles are not visible in the micrograph). Figure 9 shows thin films taken from specimens of PMMA and the modified ductile PMMA with 47 wt% 3 layer particles and deformed at different speeds. The fibrillar structure of the cavitated particle shells is visible in each case, and crazing is fully suppressed in the more ductile matrix at 0.1 mm/s (cf. Figure 7). At high v the tendency to craze increases, with numerous "short sharp" crazes emanating from the modifier particles in both materials, although deformation remains highly delocalized, consistent with the ductile impact response of the corresponding bulk specimens.

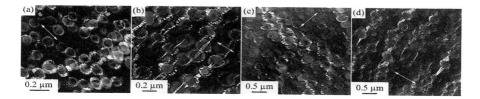

Figure 9.   Thin films of PMMA containing 47 wt% 3 layer particles deformed at (a) 0.1 mm/s and (b) 1 m/s and chemically modified PMMA (cf. Figure 7) containing 47 wt% 3 layer particles deformed at (c) 0.1 mm/s and (c) 1 m/s [37] (tensile direction indicated by the arrows).

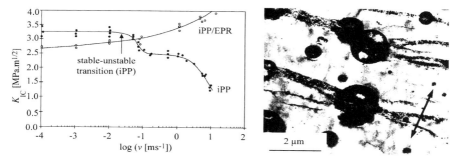

Figure 10. K_{IC} as a function of v in iPP with M_w = 455 kg/mol and iPP with M_w = 435 kg/mol containing 15 wt% EPR; the image shows modifier particle cavitation and crazing in iPP/EPR deformed at about 1 m/s and stained with RuO_4 [38].

Isotactic polypropylene

Unlike commercial grades of PMMA, iPP is often relatively crack resistant at low v. However, its impact properties remain a limiting factor in applications, owing to ductile-brittle transitions such as that shown in Figure 10 [38]. These reflect an increased tendency to craze at high v and low T for fixed M, as in PMMA. The ductile-brittle transitions are also relatively sensitive to M in the range corresponding to commercial materials. Thus, ductile-brittle transitions at fixed T and v are also induced when M is reduced (e.g. by thermal degradation [39]). A craze microstructure typical of bulk iPP is shown in Figure 11a, with fibril diameters comparable with those in crazes in glassy polymers (around 10 nm [40]). Extensive interlamellar cavitation may also occur in the early stages of deformation as a precursor to craze formation. This is conveniently observed using thin films (Figure 11b), although the lack of plastic constraint in these latter favours diffuse lamellar shear (Figure 11c), rather than the development of well-defined craze morphology, as deformation proceeds. Given that typical modifier particle diameters (see below) are also generally much greater than the thin film thicknesses required for TEM, thin film techniques are therefore of somewhat limited relevance to bulk deformation in this case.

In bulk fracture specimens, ductile behaviour is generally characterised by stress whitening and the formation of a wedge-shaped deformation zone of multiple crazing and/or shear deformation at the crack tip. However, at impact speeds, a single narrow craze is observed, propagating normal to the principal stress axis, as shown in Figure 12, which implies crazing in this regime to be insensitive to the local morphology [17].

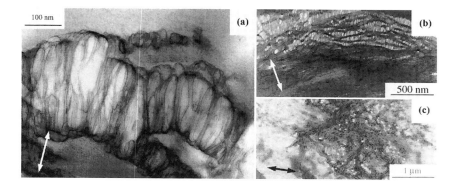

Figure 11.   (a) Fibrillar deformation in bulk α iPP deformed at room temperature, embedded in PMMA and stained with RuO_4 vapour; (b) early stages of irreversible deformation in a thin film of αiPP deformed at room temperature and post-stained with RuO_4 vapour, showing interlamellar cavitation; (c) diffuse lamellar shear in the same film [38].

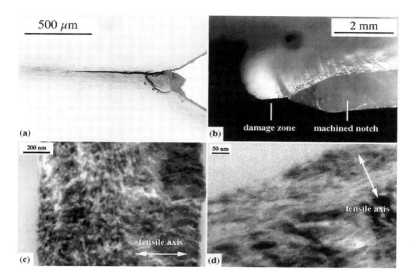

Figure 12. (a) Side-view of the crack-tip damage zone in a CT specimen of iPP with M_w of 455 kg mol^{-1} deformed at about 3 ms^{-1}; (b) oblique view of the damage zone showing the curved deformation front; (c) TEM micrograph of the collapsed fibrillar structure of the crack-tip craze; (d) detail of structure at the craze-bulk interface [18].

Influence of matrix structure: The tendency of iPP to show crazing and brittle behaviour may be linked at least partly to the unusual "cross-hatched" morphology of the relatively stable monoclinic α modification, which dominates in conventional grades under standard processing conditions [41, 42]. This morphology may block certain slip mechanisms [43, 44], and hence not only raise the yield stress but also accentuate the yield drop, thus favouring localised deformation modes such as crazing. The fracture behaviour of iPP is dependent on many other factors, large spherulites promoting brittleness, for example, owing to the concentration of structural defects and impurities at their boundaries [45]. It is nevertheless widely held, mainly on the basis of impact data, that toughness can be improved by preferential nucleation of the β phase, whose spherulites have a more conventional morphology [46-48]. Certainly, in CT tests there is a significant increase in fracture toughness in the presence of the β phase in certain ranges of T and

v, and an increase in the ductile-brittle transition speed by a least three decades at $T > T_g$ has been reported [21]. Figure 13 shows BSE-SEM images of crack-tip deformation in a β nucleated CT specimen in the ductile regime, confirming deformation to be relatively delocalized, and strongly correlated with the spherulitic structure. That crazes in α iPP are more localised and better defined than in β iPP may therefore reflect not only the influence of the cross-hatched structure on lamellar slip described in the previous section, but also the relatively homogeneous lamellar textures of α iPP spherulites. It is nevertheless difficult to exclude other factors, such as crystallinity and spherulite size, as well as the intrinsically higher molecular mobility of β iPP, purely on the basis of such microscopic observations [21].

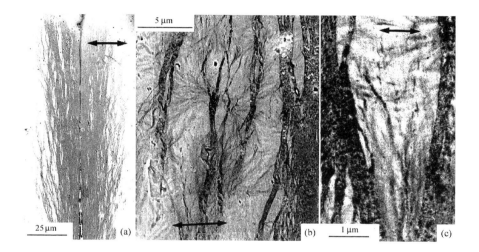

Figure 13. BSE-SEM images of crack-tip microdeformation in iPP containing approximately 80 wt% β phase tested at 0.001 ms^{-1} and 25 °C and stained with RuO$_4$: (i) overview of the crack tip damage zone; (ii) and (iii) details from the periphery of the damage zone. The arrows indicate the loading direction [21].

Rubber toughened iPP: Rubber toughened grades of iPP are available with a wide range of formulations and microstructures. The most common modifiers are ethylene propylene rubber (EPR), often introduced by varying the composition during synthesis (reactor blending) and ethylene-propylene-diene monomer elastomer (EPDM) [49]. A corresponding diversity of factors contribute to mechanical performance (modifier T_g and content, particle size and size distribution, modifier morphology, interfacial strength, matrix plasticization, crystallinity and phase behaviour, and matrix molecular architecture [50-56]). Although the particles are introduced by blending rather than as prefabricated entities with precise morphologies, loss in stiffness at high modifier contents can again be moderated by using modifiers that adopt core-shell morphologies, such as shown in Figure 3, where the particle cores consist of ethylene-rich semicrystalline inclusions.

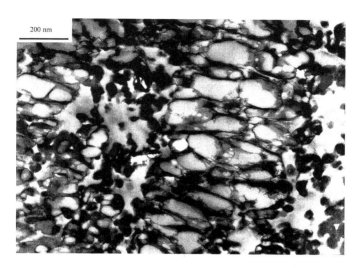

Figure 14. TEM of deformation close to the crack tip in a bulk specimen of iPP containing 52 vol% EPDM (RuO$_4$ stained, tensile axis roughly horizontal) [58].

As in PMMA, a primary function of rubber toughening is to delocalize crack-tip deformation; the stiffness mismatch between the matrix and modifier leads to local stress concentrations at the particle equators, promoting nucleation of crazes or precursor regions of interlamellar cavitation, as shown in Figure 10, for example. The accompanying plastic constraint release may then favour ductile drawing of the interparticle ligaments. Moreover, at high discrete particle contents, achievable in iPP toughened with EPDM vulcanates, for example, constraint release due to particle cavitation is sufficient for crazing to be entirely replaced by ductile drawing of the interparticle ligaments over a wide range of conditions, as described previously for PMMA. In the example shown in Figure 14, cavitation occurred along rows of particles aligned roughly perpendicular to the tensile axis, sometimes referred to as "croids" [57], the subsequent necking of the intervening matrix ligaments leading to structures analogous to crazes, albeit at very different length scales [58].

Matrix properties and their dependence on test conditions are also important for the performance of toughened iPP [59, 60]. The effect of M closely follows that of untoughened iPP, with low M favouring crazing and brittle behaviour, and high M favouring shear deformation. Crazing is also favoured over homogeneous deformation in the toughened materials at high speeds and low temperatures, although this does not necessarily lead to brittle behaviour [61]. Indeed, toughening, as reflected by improvements in K_{IC}, say, is most marked in regimes of v in which the untoughened matrix is brittle, i.e. in regimes in which crazing dominates. In the example shown in Figure 10, the toughness continues to increase with increasing v beyond the ductile-brittle transition of the untoughened matrix, although at the highest v the behaviour becomes difficult to interpret, owing to significant adiabatic heating at the crack-tip. (A 90 K rise in temperature has been reported for toughened iPP at speeds of 10 ms^{-1} [54].) Moreover, β nucleation has little effect on the brittle-ductile transition of rubber-toughened iPP at any temperature. Given the important role of the rubber particles in initiating matrix deformation during crack initiation and propagation, this supports the earlier suggestion that improvements in toughness of the homopolymer on β nucleation are mainly morphological in origin [21].

Conclusions

Different morphologies and/or modifier contents can lead to surprisingly similar fracture behaviour in bulk PMMA, although the microdeformation behaviour, as revealed by TEM and LV-SEM, varies considerably. Widespread crazing of the matrix associated with cavitation of the rubbery phase is generally associated with ductile behaviour in PMMA with intermediate spherical modifier particle contents. However, results for IPNs and high spherical particle contents indicate that crazing is not always essential for toughening. The fundamental requirement is suggested to be the establishment of a continuous network of cavities and low shear modulus material (rubber or crazed material), able to release triaxial constraints on shear deformation. There are strong parallels between the microdeformation behaviour of toughened iPP and PMMA in this respect, crazing becomes limited in both materials as the modifier content approaches 50 wt.-%. Moreover, there is a trend to more brittle behaviour with increasing test speed in both cases, which may be accounted for in terms of the time-dependence of damage development. Transitions to unstable crack propagation, i.e. macroscopically brittle behaviour, may be seen as a response of the material over timescales in which delocalized deformation in the matrix is kinetically limited, as reflected by the observation of a single crack-tip craze in bulk specimens, or isolated crazes in thin films of toughened PMMA. Indeed, in certain toughened PMMAs, thin film studies are particularly useful for identifying such trends in the microdeformation behaviour, and are therefore of direct relevance to bulk fracture, whilst at the same time giving access to a wealth of local detail that the staining techniques necessary for bulk specimens may obscure.

Acknowledgments

We acknowledge the generous financial support of TotalFinaElf, Ciba Speciality Chemicals and the Swiss CTI during the course of this work, and the technical support of the Electron Microscopy Centre (CIME) of the EPFL.

[1]   P. Béguelin, H.-H. Kausch, in *"Impact and Dynamic Fracture of Polymers and Composites"*, J. G. Williams, A. Pavan, Eds., Mechanical Engineering Publications, London 1995.
[2]   E. J. Kramer, *Adv. Polym. Sci.* **1983**, *52-53*, 1.
[3]   H. R. Brown, *Macromolecules* **1991**, *24*, 2752.
[4]   C. B. Bucknall, *"Toughened Plastics"*, Appl. Sci. Publishers, London 1977.
[5]   C. K. Riew, A. J. Kinloch, Eds., *"Toughening of Plastics II"*, ACS Symposium Series 252, ACS, Washington DC 1996.
[6]   R. A. Pearson, H.-J. Sue, A. F. Yee, Eds., *"Toughening of Plastics: Advances in Modelling and Experiments"*, ACS Symposium Series 759, ACS, Washington DC 2000.
[7]   A. M. L. Magalhães, R. J. M. Borggreve, *Macromolecules* **1995**, *28*, 5841.
[8]   C. J. G. Plummer, P. Béguelin, H.-H. Kausch, *Colloids and Surfaces A* **1999**, *153*, 551.
[9]   Z. Bartczack, A. S. Argon, R. E. Cohen, M. Weinberg, *Polymer* **1999**, *40*, 2331.
[10]   Z. Bartczack, A. S. Argon, R. E. Cohen, M. Weinberg, *Polymer* **1999**, *40*, 2347.
[11]   R. J. M. Borggreve, R. J. Gaymans, J. Schuijer, J. F. Ingen Housz, *Polymer* **1989**, *28*, 1489.
[12]   S. Wu, *Polymer* **1985**, *26*, 1855.
[13]   S. Wu, *J. Appl. Poly. Sci.* **1988**, *35*, 549.
[14]   P. Béguelin, PhD Thesis, EPFL, 1996.
[15]   P. Béguelin, C. Fond, H.-H. Kausch, *Int. J. Fracture* **1998**, *89*, 85.
[16]   J.G. Williams, *"Fracture Mechanics of Polymers"*, Ellis Horwood, Chichister UK 1984.
[17]   C.J.G. Plummer, P. Scaramuzzino, H.-H. Kausch, *Polym. Eng. & Sci.* **2000**, *40*, 1306.

114

[18] R. Gensler, C. J. G. Plummer, C. Grein, H.-H. Kausch, *Polymer* **2000**, *41*, 3809.
[19] K. Friedrich, *Adv. Polym. Sci.* **1983**, *52-53*, 225.
[20] H.B.H. Hamouda, M. Simaos-Betbeder, F. Grillon, P. Blouet , N. Billon, R. Piques, *Polymer* **2001**, *42*, 5425.
[21] C. Grein, C. J. G. Plummer, H.-H. Kausch, Y. Germain, P. Béguelin, *Polymer* **2002**, *43*, 3279.
[22] J. H. Butler, D. C. Joy, G. F. Bradley, S. J. Krause, *Polymer* **1995**, *36*, 1781.
[23] G. Binning, C. F. Quate, C. Gerber, *Phys. Rev. Lett.* **1986**, *56*, 930.
[24] B. D. Lauterwasser, E. J. Kramer, *Phil. Mag.* **1979**, *39A*, 469.
[25] C. J. G. Plummer, H.-H. Kausch, *Macromol. Chem. Phys.* **1996**, *197*, 2047.
[26] C. J. G. Plummer, H.-H. Kausch, *J. Macromol. Sci. - Phys.* **1996**, *B35*, 637.
[27] G. H. Michler, *Trends in Polymer Science* **1995**, *3*, 124.
[28] C. J. G. Plummer, P. Béguelin, H.-H. Kausch, *Polymer* **1996**, *37*, 7.
[29] C. J. G. Plummer, L. Tézé, J. L. Halary, L. Monnerie, H.-H. Kausch, *Polymer* **1996**, *37*, 19.
[30] P. A. Lovell, *Trends in Polymer Science* **1996**, *4*, 264.
[31] P. A. Lovell, J. McDonald, D. E. J. Saunders, M. N. Sherratt, R. J. Young, in *"Toughened Plastics I"*, C. K. Riew, A. J. Kinloch, Eds., ACS, Washington DC 1993.
[32] P. A. Lovell, M. M. Sherratt, R. J. Young, in *"Toughened Plastics II"*, C. K. Riew, A. J. Kinloch, Eds., ACS, Washington DC 1996.
[33] C. J. Hooley, D. R. Moore, M. Whale, M. J. Williams, *Plast. Rubber Processing & Applications* **1981**, *1*, 345.
[34] O. Julien, P. Béguelin, L. Monnerie, H.-H. Kausch, in *"Toughened Plastics II"*, C.K. Riew, A.J. Kinloch, Eds. ACS, Washington DC 1996.
[35] P. Béguelin, C. J. G. Plummer, H.-H. Kausch, in *"Polymer Blends and Alloys"*, G. O. Shonaike, G. P. Simon, Eds., Marcel Dekker, New York 1999.
[36] P. Heim, C. Wrotecki, M. Avenel, P. Gaillard, *Polymer* **1993**, *34*, 1653.
[37] L. Dupuits, C. J. G. Plummer, P. Gerard, J.-A. E. Månson, *Proc. Int. Conf. on Deformation, Yield and Fracture of Polymers*, Cambridge, 7-10 April 2003, p. 399.
[38] R. Gensler, PhD Thesis, EPFL, 1998.
[39] R. Gensler, C. J. G. Plummer, H.-H. Kausch, E. Kramer, J.-R. Pauquet, H. Zweifel, *Polym. Degrad. & Stab.* **2001**, *67*, 195.
[40] C. J. G. Plummer, C. Creton, F. Kalb, L. Léger, *Macromolecules* **1998**, *31*, 6164.
[41] R. H. Olley, D. C. Bassett, *Polymer* **1989**, *30*, 399.
[42] D. R. Norton, A. Keller, *Polymer* **1985**, *26*, 704.
[43] M. Aboulfaraj, C. G'Sell, B. Ulrich, A. Dahoun, *Polymer* **1995**, *36*, 731.
[44] G. Coulon, G. Castelein, C. G'Sell, *Polymer* **1999**, *40*, 95.
[45] K. Friedrich, U. A. Karsch, *J. Mater. Sci.* **1981**, *16*, 2167.
[46] J. Karger-Kocsis, J. Varga, G. W. Ehrenstein, *J. Appl. Polym. Sci.* **1997**, *64*, 2057.
[47] S. C. Tjong, J. S. Shen, R. K. Y. Li, *Scripta Metallurgica et Materialia* **1995**, *33*, 503.
[48] M. W. Gahleitner, *J. Appl. Polym. Sci.* **1996**, *61*, 649.
[49] J. Karger-Kocsis, Ed. *"Polypropylene - an A to Z reference"*, Kluwer, Dodrecht 1999.
[50] D. Dompas, G. Groeninckx, M. Isogawa, T. Hasegawa, M. Kadokura, *Polymer Commun.* **1995**, *36*, 437.
[51] A. van der Wal, R. Nijhof, R. J. Gaymans, *Polymer* **1999**, *40*, 6031.
[52] A. van der Wal, A. J. J. Verheul, R. J. Gaymans, *Polymer* **1999**, *40*, 6067.
[53] A. van der Wal, A. J. J. Verheul, R. J. Gaymans, *Polymer* **1999**, *40*, 6057.
[54] A. van der Wal, R. J. Gaymans, *Polymer* **1999**, *40*, 6045.
[55] B. Dukanszky, F. Tudos, A. Kallo, G. Bodor, *Polymer* **1989**, *30*, 1399.
[56] F. Ramsteiner, *Acta Polymerica* **1991**, *42*, 584.
[57] H.-J. Sue, *J. Mater. Sci.* **1992**, *27*, 3098.
[58] F. Kalb, L. Léger, C. J. G. Plummer, C. Creton, P. Marcus, A. M. L. Magalhaes, *Macromolecules* **2001**, *34*, 2702.
[59] A. van der Wal, J. J. Mulder, R. J. Gaymans, *Polymer* **1998**, *39*, 5467.
[60] A. van der Wal, J. J. Mulder, R. J. Gaymans, *Polymer* **1998**, *39*, 5477.
[61] B. Z. Jang, D. R. Uhlmann, J. B. Van der Sande, *J. Appl. Polym. Sci.* **1985** *30*, 2485.

Morphology and Properties of Particulate Filled Polymers

Béla Pukánszky,[*1,2] *János Móczó*[1,2]

[1]Research Laboratory of Materials and Environmental Chemistry, Hungarian Academy of Sciences, H-1525 Budapest, P.O. Box 17, Hungary
[2]Department of Plastics and Rubber Technology, Budapest University of Technology and Economics, H-1521 Budapest, P.O. Box 92, Hungary

Summary: Although the structure of particulate filled polymers is usually thought to be very simple, often structure related phenomena determine their properties. Segregation occurs only when long flow paths and large particles are used in production. The occurrence and extent of aggregation depend on the relative magnitude of attractive and separating forces, which prevail during the homogenization of the composite; the balance of adhesive and shear forces determines structure. Fillers of small particle size always aggregate, usually leading to decreased strength and especially low impact resistance. Anisotropic particles (talc, mica, short fibers) are orientated during processing. ESR is a relatively simple technique for the estimation of orientation and orientation distribution, which are determined by processing conditions, i.e. flow pattern, shear conditions, mold filling rates, cooling conditions, etc. The orientation of the particles strongly affects composite stiffness and strength. In practice, often several factors simultaneously influence the properties of products prepared from particulate filled polymers. Separation of the effects of the influencing factors is difficult, although such knowledge would help to control composite properties. The structure and properties of injection and compression moulded PP composites containing $CaCO_3$ or talc differs considerably from each other. The aggregation of $CaCO_3$, the nucleating effect and the orientation of talc affect product properties. The latter are also influenced by the skin-core structure developing during injection molding as well as by the orientation of the polymer. An example is discussed in this paper, which facilitates the identification of the effect of these factors with the help of a simple model and indicates a way in which product properties can be controlled.

Keywords: aggregation; crystalline structure; mechanical properties; orientation; particulate fillers; segregation; structure-property correlation

Introduction

The structure of semi-crystalline polymers, including polypropylene (PP), is relatively complicated. They can crystallize in different modifications. Their crystallinity and size, as well as size distribution of the crystalline units (lamellae, spherulites) vary over a wide range with changing sample preparation and processing conditions[1-4]. The introduction of a second component into PP usually further modifies morphology. Fillers may act as nucleating agents, changing thermodynamic and kinetic conditions of crystallization[5-8]. Crystal modification, the size of crystalline units and the amount of the crystalline phase can all change as a result[9]. The introduction of particulate fillers often leads to the development of specific morphology; anisotropic particles are usually orientated to different extents[10,11], while spherical fillers frequently form aggregates[12,13]

Because of the considerable number of factors influencing the structure and properties of particulate filled polypropylene, the views concerning their effects are often contradictory. In some cases the direct effect of a certain aspect of crystalline morphology is claimed to determine properties, in others such an influence is denied completely. Hutley and Darlington[14,15], for example, found a direct correlation between the crystallization temperature and impact strength of particulate filled PP, while Maiti et al.[16] observed linear dependence of some tensile characteristics on crystallinity. Kendall[17], on the other hand, denied the effect of matrix morphology on composite properties and emphasized the importance of interfacial interaction between the matrix and the dispersed phase.

Talc and $CaCO_3$ are used in large quantities for the modification of PP[18-20]. The two fillers have different particle geometry and a dissimilar effect on the crystalline structure of PP. $CaCO_3$ consists of more or less spherical particles and influences crystallization only slightly, while talc has plate-like geometry and a strong nucleation effect[9]. The properties of the composites prepared with the two fillers are also different; however, it is still debated whether the changing crystalline structure of the matrix, nucleation, or anisotropy and orientation of the filler cause the observed differences[21]. The goal of this paper is to give an overview of the factors influencing the properties of particulate filled polymers with special attention to the correlation of structure and properties. Structure-related phenomena occurring in particulate filled polymers are summarized briefly first, then a case study is presented, which shows the effect of various factors acting simultaneously in particulate filled PP and indicating a way how to separate these effects.

Segregation

The segregation of a second phase during processing is observed in some heterogeneous polymer systems[22,23]. Kubát and Szalánczi[22] investigated the separation of phases during the injection molding of polyethylene and polyamide using the spiral test. The two polymers contained large glass spheres of 50-100 μm particle size and extremely long flow paths of up to 1.6 m. They found that considerable segregation took place along the flow path; the glass content of a composite containing 25 wt% filler exceeded 40 % locally, at the end of the mold. Occasionally segregation was also observed across the cross-section of the sample; the local amount of filler was higher in the core than at the walls. Segregation depended on filler content and it became more pronounced with increasing size of the particles. Karger and Csikai[23] observed the segregation of the dispersed phase in elastomer modified PP. They also found an increased amount of the dispersed phase in the center of the specimen.

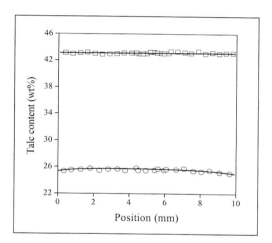

Figure 1.   Distribution of talc across the width of injection moulded PP specimens. (o) 10 vol%, (□) 20 vol% talc content.

The possible segregation of talc particles dispersed in a PP matrix was investigated in injection-moulded specimens of 4 x 10 x 150 mm dimensions. Filler content was measured across the cross-section of the specimens by thermogravimetric analysis. As Figure 1 shows, no differences in filler content were detected as a function of position; the particles were homogeneously distributed in the PP matrix, independent of the average filler content. These and other results indicate that under practical conditions (small particles, relatively high filler content, normal flow path) segregation is of secondary importance, and we may assume the homogeneous distribution of the dispersed phase in the matrix polymer.

Aggregation

Particulate filled composites are produced almost exclusively by the melt mixing of the components, when the shear stresses developing in the processing machine try to separate particles attached to each other[24]. The occurrence and extent of aggregation depend on the relative magnitude of adhesion and separating forces[25-27]. The former is determined by the surface tension of the filler and its particle size, while the latter also depends on particle size as well as on the level of shear forces. As a consequence, aggregation may be decreased by surface treatment, by increasing the particle size of the filler or shear forces[25]. Commercial grades of $CaCO_3$ usually have a wide particle size distribution, thus a fraction of the small particles always aggregates, while large particles are distributed separately. The unambiguous determination of aggregation is difficult. Various techniques can be used, including measurements carried out on the dry filler, in suspension or on the composite itself[28-33].

Previous results have shown that aggregation always occurs below a certain particle size or above a certain specific surface area[26,27]. In PP composites, the critical value was 5-7 m^2/g. Aggregation modifies stiffness only slightly[26], but strength and impact resistance depend very much on structure, both decrease with increasing extent of aggregation[26,27]. The mode of failure initiation depends also on particle size. Debonding is the dominat deformation mechanism in composites containing large particles, while cracks are initiated inside aggregates forming at smaller particles sizes[27]. Contradictory results were obtained on the effect of processing technology; injection-moulded specimens did not appear to be always more homogeneous than compression moulded ones[27]. However, a detailed analysis of actual processing conditions and the determination of the number of aggregates proved that properties can be related to the extent of aggregation[34]. On the other hand, deviations from the general tendency indicated that some factors also influence properties, which had not been taken into account during the analysis.

One of these factors is the strength of the aggregates. The strength of agglomerates is of considerable importance in a number of industries, i.e. in the granulating of powders, briquetting, pelletizing, etc. As a consequence, several theories were developed for the prediction of agglomerate strength. Manas-Zloczower[35] gives a brief overview of them in a monograph focusing on the mixing and compounding of polymers. Rumpf[36] proposed one of the first equations for the calculation of the tensile strength of aggregates:

$$T = \frac{9}{8} \frac{1-\varepsilon}{\pi d^2} c F \tag{1}$$

where T is the tensile strength of the aggregate, d is the diameter of the spherical particles in it, ε is the void volume fraction within the total volume, c is the mean coordination number and F is the mean interparticle force. According to the model the tensile strength of aggregates depends mainly on the particle size of the filler, and it increases with decreasing particle diameter. Cheng[37,38] suggested a somewhat different relationship for T:

$$T = \frac{3}{4} H^0 \frac{\bar{d}\,\bar{s}}{\bar{v}} \frac{\rho/\rho_s}{1-\rho/\rho_s} F_{pp}^0 \tag{2}$$

where H^0 is the effective surface separation distance of zero tensile strength; $\bar{d}$, $\bar{s}$, and $\bar{v}$ are the mean diameter, surface area and volume of the particles, ρ is bulk density and ρ_s the density of the particles, while F_{pp}^0 is the interparticle force per unit fracture area. Kendall[39] strongly criticized the approach of Rumpf[36] and followed a completely different approach; he proposed a model for aggregate strength, which is based on fracture mechanics:

$$T = 15.6 \, \phi^4 \, \Gamma_c^{5/6} \, \Gamma^{1/6} (d \, c)^{-1/2} \tag{3}$$

where ϕ is the volume fraction of the filler, Γ_c the fracture energy of the agglomerate, Γ the interfacial energy between the particles with a diameter of d, and c is the length of a macroscopic flaw in the aggregate. The validity of all three models was checked on agglomerated particulate material and good agreement was found between prediction and the experimental results obtained.

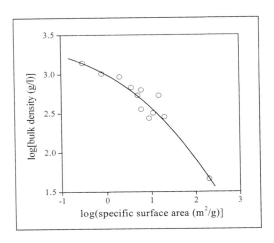

Figure 2. Correlation of the bulk density and specific surface area of various commercial CaCO₃ fillers.

Although Manas-Zloczower[35] suggests the use of the same or similar correlation for the prediction of aggregate strength in a polymer matrix, very limited experimental evidence exists to support such an approach. Nevertheless, we may consider the effect of some of the parameters included in Eqs. 1 and 2 on aggregate strength. The bulk density of the filler can be measured and, according to Eq. 2, the tensile strength of the aggregates should become smaller with decreasing bulk density. In Figure 2, the bulk densities of 11 CaCO₃ fillers[26] with a wide range of particle characteristics are plotted against their specific surface areas on a logarithmic scale. We can see that with increasing surface area, i.e. decreasing particle size, the bulk density of the filler decreases significantly, indicating strongly decreasing tensile strength of the aggregates. The other important parameter, which appears in all equations, is the size of the filler particles. According to Eq. 1, the tensile strength of the aggregates is proportional to the reciprocal square of particle diameter. We plotted this quantity as a function of the specific surface area of the fillers in Figure 3. The $1/d^2$ value covers more then 10 orders of magnitude, which indicates that this parameter has indeed a strong influence on aggregate strength. If Eq. 1 is valid, the tensile strength of the aggregates should increase drastically with decreasing particle size.

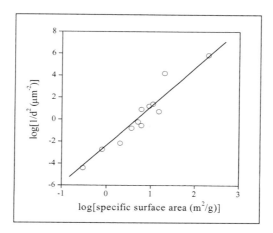

Figure 3.   Dependence of aggregate strength on the particle size of the filler for various $CaCO_3$ fillers.

Unfortunately we could not check the effect of bulk density and particle size on aggregate strength directly, but we could obtain some information about it from the analysis of composite properties. The tensile strength of PP composites is plotted against filler content in Figure 4. We know that under the effect of external load large particles easily debond from the matrix, which leads to a continuous decrease in tensile strength with increasing filler content. With decreasing particle size the effect of interfacial interactions become more important and tensile strength increases as an effect of the formation of a hard interlayer[25].

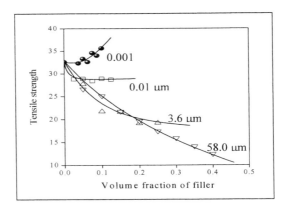

Figure 4.   Effect of filler particle size and composition on the tensile strength of PP composites.

At particle sizes smaller than 1.0 μm aggregation becomes noticeable, but the strength of the aggregates is low[27]. However, the tensile strength of composites containing very small particles increases with filler content, indicating that debonding is no longer the dominating deformation mechanism. Moreover, the bulk density of this filler is 46 g/l, compared to the ~1400 g/l of the largest particles, which proves that Eq. 1 predicts aggregate properties better than Eq. 2 does. Although aggregate characteristics must still be studied in more detail in the future, aggregate strength may indeed influence composite properties. We must mention here, though, that PP composites containing 0.001 μm large particles are very stiff and brittle, thus their practical relevance is relatively small.

Orientation

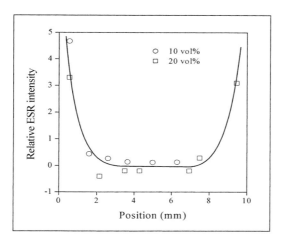

Figure 5. Orientation of talc particles across specimen thickness in injection moulded PP composites of different filler contents.

The orientation of anisotropic filler particles occurs in all processing operations, and orientation, as well as orientation distribution was shown to drastically influence composite properties. Orientation in extruded and compression-moulded products is relatively homogeneous, while it changes across the thickness and width of injection moulded parts according to the flow pattern developing during mold filling. Several methods are available for the determination of particle orientation, but most of them are rather tedious. ESR spectroscopy is a relatively simple technique which can be used for the determination of local and average orientation[40,41]. It uses the signal of transition metal contamination found in every mineral for the determination of orientation. The shape and the intensity of the signal depend on the relative orientation of the crystal planes and the magnetic field. A comparison of the spectrum of the powder and that of the oriented specimens offers a way to estimate the relative orientation of the particles. Further details of the technique can be found elsewhere[40,41]. Figure 5 shows the orientation distribution of talc particles in a specimen of 4 x 10 mm cross-section, injection moulded from a PP/talc composite of 20 vol% filler content. The distribution agrees well with experience and shows strong parallel orientation of the particles to flow direction at the wall of the mold and their random orientation in the core of the specimen. The strength and stiffness of the composites increase considerably, parallel to the orientation of anisotropic particles, and decrease perpendicular to that[42].

Structure-property Correlations

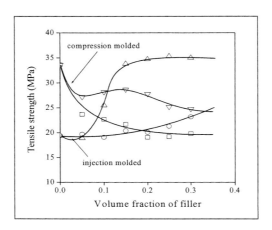

Figure 6.   The effect of various parameters on the tensile strength of PP composites; ($\Box$, $\bigcirc$) CaCO$_3$, ($\bigtriangledown$, $\triangle$) talc.

The effects of filler contents on the tensile strength of particulate filled PP composites are plotted in Figure 6. The composites were injection or compression moulded, and contained CaCO$_3$ or talc particles, i.e. three factors were changed in this series of experiments: filler type, composition and processing technology. The strengths of the various composites differ significantly from each other; that of the PP/talc composites of 30 vol% filler content exceeds by far the strength of the material containing CaCO$_3$ in the same amount. As discussed before, several structure related factors may influence the properties of these composites, some are associated with the filler and some with the matrix polymer. The crystal modification of the polymer, the size of the morphological units, crystallinity, as well as the orientation of the crystalline and amorphous phases, all affect the properties of the product. Moreover, all are influenced by processing technology and modified by the strong nucleating effect of talc. The distribution of the filler, homogeneity, aggregation, the anisotropy of filler particles and their orientation, as well as orientation distribution, may also play an important role in the determination of composite properties. Naturally more than one factor acts simultaneously and the determination of the dominating one is difficult. The possible effects of the various factors are analyzed quantitatively in the following sections and an attempt is made to determine the most important ones.

Experimental

Stamylan P16M10 (DSM, The Netherlands) polypropylene homopolymer was used as matrix polymer. Hydrocarb 95 T CaCO$_3$ (Omya, Switzerland – A$_f$ = 11.0 m^2/g, d = 1.0 μm) and Luzenac 10M00S talc (Luzenac, France – A$_f$ = 8.0 m^2/g, d = 3.4 μm) were added to PP to produce the composites processed by injection molding.

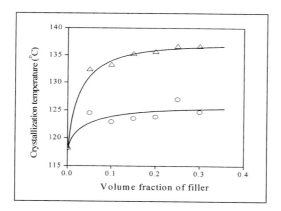

Figure 7.   Effect of filler type and content on the crystallization temperature (lamella thickness[43,44]) of the PP matrix; (○) CaCO$_3$, (△) talc.

Different grades of filler were introduced into the composites prepared by compression molding:

Durcal 2 (Omya, Switzerland – A_f = 3.3 m^2/g, d = 3.6 μm) and Finntalc M05 (Finnminerals, Finland – A_f = 8.4 m^2/g, d = 2.8 μm). The composition changed between 0 and 0.3 volume fraction in 0.05 volume-fraction steps. Homogenization was carried out using a Brabender DSK 42/7 twin-screw compounder; the blends and composites were cooled in a water bath and pelletized. Specimens were prepared by compression or injection molding, respectively. Tensile testing was carried out at 50 mm/min cross-head speed. Young's modulus (E), yield (σ_y, ε) and ultimate (σ, ε) properties were determined from recorded force vs. elongation traces. The nucleation effect of the fillers and the crystallinity of the composites were studied in the DSC 30 cell of a Mettler TA 3000 Thermal Analysis System. 10 mg samples were measured in two heating and cooling cycles, at a rate of 10 °C/min. Relative changes in filler orientation were followed by ESR spectroscopy[21,40,41]. The dispersed structure and orientation of the talc particles were studied by scanning electron microscopy (SEM) on fracture surfaces prepared at liquid nitrogen temperature.

Crystalline Structure

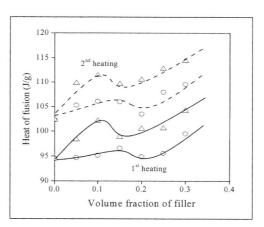

Figure 8.   Influence of processing technology and filler type as well as content on the crystallinity of the matrix polymer. Symbols are the same as in Fig. 7.

The nucleation effect of a filler can be deduced from non-isothermal crystallization experiments;

the onset of crystallization as well as the crystallization peak temperature are related to nucleation and thus to the size of the crystalline units[43,44]. In Fig. 7 the crystallization peak temperatures of the composites are plotted as a function of composition for injection moulded composites

According to this figure, talc has a strong nucleating effect while $CaCO_3$ a much weaker one. The different effects on nucleation and the resulting dissimilarities in crystalline morphology may be the cause of the differences in the properties of talc and $CaCO_3$ filled PP composites. The crystallinity of the composites prepared with the two fillers shows characteristic differences too (Figure 8). Crystallinities measured in the first and the second heating cycles, respectively, differ considerably from each other. The first cycle reflects the effect of processing technology, i.e. injection molding in this case, on crystallinity. During the first heating cycle this is completely erased and only the effect of the filler influences crystallinity in the second run. According to the figure the effect of processing technology and the introduction of the filler are completely independent from each other. The orientation of the amorphous and crystalline phases has not been measured, although these might also influence properties.

Particle Related Structure

The presence of aggregates was not detected in the composites. The shape of $CaCO_3$ and talc particles differ significantly from each other. Although $CaCO_3$ particles are not spherical, their aspect ratio is close to 1. The anisotropy of talc particles is more significant; their aspect ratio is around 20-30. Although contradictory information is reported on the effect of anisotropy, it certainly influences composite properties. During sample preparation, i.e. processing, orientation of the particles takes place even under the apparently mild shearing conditions of compression moulding[21,40]. Both $CaCO_3$ and talc orientates, but the final result is evidently different because of the different aspect ratio of the two fillers.

Average values of orientation were measured both in injection moulded specimens and compression moulded plates by ESR spectroscopy. Relative values related to the sample with the smallest orientation are plotted in Figure 9 as a function of composition. Composition dependence differs, it decreases continuously in the compression moulded specimens and exhibits a maximum in the injection moulded specimens. The tendency in the composition dependence of orientation indicates that with increasing filler content filler particles hinder each other in movement, and the probability of particle interaction increases. Not only is the average orientation different, but also its distribution in the specimens prepared by the two techniques. Most of the talc particles are oriented normal to the direction of pressure in the compression moulded plate, while the complicated flow pattern of injection molding leads to a different distribution. Particles are oriented parallel to the wall near to it and almost vertically to that in the core of the specimen. It is obvious from these results that the orientation of anisotropic filler particles changes significantly strongly with composition and that these changes may have a pronounced effect on properties.

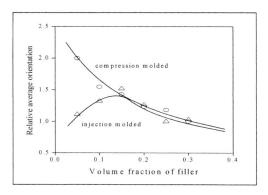

Figure 9.   Effect of processing technology on the orientation of talc particles in PP composites. Symbols are the same as in Fig. 7.

Quantitative Evaluation

According to Figure 6 the strength of the composites changes both with processing technology and filler type. Talc reinforces PP relative to $CaCO_3$ and causes even absolute reinforcement in injection moulded samples. A decrease in strength is observed as a function of filler content for the compression moulded specimens. The composition dependence of strength can be predicted by a simple model, which takes into account the effective load-bearing cross-section of the matrix, the strain hardening of the polymer due to elongation and matrix/filler interaction[45]:

$$\sigma_T = \sigma_{T0}\, \lambda^n\, \frac{1-\varphi}{1+2.5\varphi}\, \exp(B\,\varphi) \qquad (4)$$

where σ_T and σ_{T0} are the true tensile strengths ($\sigma_T = \sigma\lambda$, $\lambda = L/L_0$) of the composite and the matrix, respectively, n is a parameter expressing the strain hardening tendency of the matrix and B is proportional to stress transfer from the matrix to the dispersed phase, i.e. to interaction. When strength is expressed in reduced form, $\sigma_{Tred} = \sigma_T (1 + 2.5\varphi)/(1 - \varphi)$, or relative form, $\sigma_{Trel} = \sigma_{Tred}/\sigma_{T0}$, linear correlation of the variables should be obtained as a function of composition in a semi-logarithmic plot.

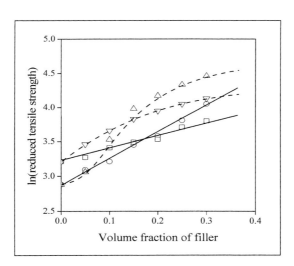

Figure 10. Reduced strength of particulate filled PP composites plotted against filler content in linear form. Symbols are the same as in Figure 6.

Figure 10 shows reduced strength in linear representation. $CaCO_3$ filled composites correspond to the prediction; they give straight lines with different slopes and intersections. Intersections represent matrix properties, which reflect the dissimilarities in structure resulting from the two processing technologies. It has been shown earlier that B depends on the size of the interface and the thickness of the interphase[46], i.e.

$$B = \left(1 + A_f \, \rho_f \, \ell\right)\ln \frac{\sigma_{Ti}}{\sigma_{T0}}$$

(5)

where A_f and ρ_f are the specific surface area and the density of the filler, while ℓ and σ_{Ti} are the thickness and strength of the interphase, respectively. The investigated $CaCO_3$ fillers differ both in specific surface area and interaction, since Hydrocarb 95 T was surface treated with stearic acid. It has been shown, however, that specific surface area has a more pronounced effect on the value of B than on surface treatment[47,48]. Indeed, composites containing the filler with the larger specific surface area exhibit a steeper slope, in spite of the weaker interfacial interaction in this system.

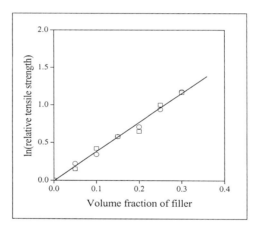

Figure 11. Relative strength of PP/CaCO₃ composites normalized by taking into account the differences in filler characteristics (B). (○) injection moulded, (□) compression moulded.

After taking into account the effect of different specific surface areas and interaction of the fillers by normalizing B, the relative strengths of the CaCO$_3$ filled composites were plotted in the linear form (Figure 11). The two sets of data fall onto the same line, proving that the effect of processing technique influences matrix properties independently of the characteristics of the fillers, a factor which was eliminated by calculating relative quantities. The result also show that the dissimilarity in the properties of the two series of composites filled with the different CaCO$_3$ fillers is caused mainly by the different specific surface area of the fillers and, to a lesser extent, by different interaction (surface treatment). The independent effect of processing technology and filler characteristics is in agreement with the results of Figure 8, which showed similar composition dependence of crystallinity in the two heating runs. Injection moulding changed the absolute amount of crystallinity, but not the tendency implemented by the presence of the filler.

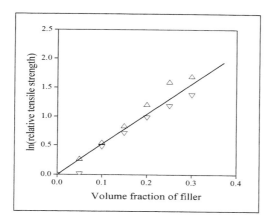

Figure 12. Linear plot of the relative tensile strength of PP/talc composites after correction for crystallinity and orientation. ($\triangle$) injection moulded, ($\triangledown$) compression moulded.

However, the non-linear character of the composition dependence of the talc filled composites still remains to be explained. In order to take the nucleation effect of the talc particles and their orientation into account, the strengths of the composites filled with talc were corrected by a simple approach. Relative crystallinity and orientation were calculated by relating strength measured at each composition to the smallest value determined in each series. Subsequently, composite strength was divided by this relative parameter characterizing structure. Reduced and relative strength values were calculated and plotted in the linear from; the latter correlation is shown in Figure 12. Correction according to crystallinity does not have any effect on strength; it does not change the shape of the correlations presented in Figure 10, it modifies their position only slightly. However, taking into account average particle orientation results in straight lines. If we plot the natural logarithm of reduced values against composition, the slope and interception of the lines obtained for two sets of composites are different again, due to the differences in processing conditions as well as in the orientation distribution of the particles. The effect of talc type can be neglected here, because the particle characteristics (specific surface area, particle size) of the two grades used are very similar to each other. These results emphasize again the major role of particle anisotropy, particle orientation and orientation distribution in the determination of composite properties.

Conclusions

The structure of particulate filled polymers is often more complicated than expected; particles are rarely distributed homogeneously in the matrix. Segregation is usually negligible under practical conditions, but aggregation must always be expected and checked for. The presence of aggregates deteriorates overall polymer properties, especially impact resistance. Anisotropic particles are always orientated, and the extent and direction of orientation determines properties. A case study on the structure and properties of particulate filled PP has shown that composites prepared with $CaCO_3$ and talc has significantly different properties. Tensile strengths were very different, both as a function of particle properties and processing condition. The application of a simple model proved that it is not the nucleating effect, but the anisometric particle geometry of talc that results in its strong reinforcing effect. Processing technology determines the structure of the matrix and the orientation of anisotropic particles; both influence properties significantly. The results obtained prove that the relation between the structure and properties of particulate filled polymers is usually quite complex, practically, always several factors influence properties simultaneously.

Acknowledgements

The authors are indebted to Prof. Antal Rockenbauer and László Korecz for the ESR measurements. Prof. Frans H.J. Maurer is acknowledged for the initiation of the project on modified PP and Dr. Erika Fekete for her continuous support during it. The help of MSc. students Judit Magyar, László Lakatos, and Péter Ács is appreciated in the study related to aggregation. The financial assistance of the National Scientific Research Fund (OTKA Grant No. 03579 and 043517) is acknowledged, for supporting the research on heterogeneous polymer systems.

[1] J.Varga, *J. Mater. Sci.* **1992**, *27*, 2557
[2] J. Varga, *J. Macromol. Sci., Phys.* **2002**, *B41*, 1121
[3] D.A. Norton, A. Keller, *Polymer* **1985**, *26*, 704
[4] J. Karger-Kocsis, *"Polypropylene: Structure, Blends and Composites"*, Elsevier, London, 1993
[5] J. Menczel, J. Varga, *J. Thermal. Anal.* **1983**, *28*, 161
[6] M. Fujiyama, T. Wakino, *J. Appl. Polym. Sci.* **1991**, *42*, 2739
[7] M. Fujiyama, T. Wakino, *J. Appl. Polym. Sci.* **1991**, *42*, 2749
[8] J. Kaiser, *Werkstofftechnik* **1990**, *80*, 330
[9] J. Varga, F. Schulek-Tóth, *Angew. Makromol. Chem.* **1991**, *188*, 11
[10] M. Fujiyama, T. Wakino, *J. Appl. Polym. Sci.* **1991**, *42*, 9
[11] T. Vu-Khanh, B. Fisa, *Polym. Compos.* **1986**, *7*, 219
[12] V. Švehlová, in: *"Polymer Composites"*, B. Sedláček, Walter de Gruyter, Berlin, 1986, p. 607
[13] V. Švehlova, E. Polouček, *Angew. Makromol. Chem.* **1987**, *153*, 197
[14] T.J. Hutley, M.W. Darlington, *Polym. Commun.* **1984**, *25*, 226
[15] T.J. Hutley, M.W. Darlington, *Polym. Commun.* **1984**, *25*, 264
[16] S.N. Maiti, P.K. Mahapatro, *Int. J. Polym. Mater.* **1990**, *14*, 205
[17] K. Kendall, *Br. Polym. J.* **1978**, *10*, 35
[18] H.P. Schlumpf, *Kunststoffe* **1983**, *73*, 511
[19] R. Rothon, *"Particulate-Filled Polymer Composites"*, Longman Sci. Techn., Harlow, 1995
[20] G. Wypych, *"Handbook of Fillers"*, ChemTec Publishing, Toronto, 1999
[21] B. Pukánszky, K. Belina, A. Rockenbauer, F.H.J. Maurer, *Composites* **1994**, *25*, 205
[22] J. Kubát, Á. Szalánczi, *Polym. Eng. Sci.* **1974**, *14*, 873
[23] J. Karger-Kocsis, I. Csikai, *Polym. Eng. Sci.* **1987**, *27*, 241
[24] P. R. Hornsby, *Adv. Polym. Sci.* **1999**, *139*, 155
[25] B. Pukánszky, E. Fekete, *Adv. Polym. Sci.* **1999**, *139*, 109
[26] B. Pukánszky, E. Fekete, *Polym. Compos.* **1998**, *6*, 313
[27] E. Fekete, Sz. Molnár, G.-M. Kim, G. H. Michler, B. Pukánszky, *J. Macromol. Sci., Phys.* **1999**, *B38*, 885
[28] S. Miyata, T. Imahashi, H. Aabuki, *J. Appl. Polym. Sci.* **1980**, *25*, 415
[29] P. R. Hornsby, C. L. Watson, *Plast. Rubber Process. Appl.* **1986**, *6*, 169
[30] Y. Suetsugu, J. L. White, *Adv. Polym. Technol.* **1987**, *7*, 427
[31] B. Bridge, M. J. Folkes, H. Jahankhani, *J. Mater. Sci.* **1989**, *24*, 1479
[32] J. C. Krapez, P. Cielo, X. Maldague, L. A. Utracki, *Polym. Compos.* **1987**, *8*, 396
[33] J. W. Ess, P. R. Hornsby, *Polym. Testing* **1986**, *6*, 205
[34] J. Móczó, E. Fekete, K. László, B. Pukánszky, *Macromol. Symp.* Accepted

134

[35] I. Manas-Zloczower, in: *"Mixing and Compounding of Polymers. Theory and Practice"*, I. Manas-Zloczower, Z. Tadmor, Eds., Hanser, Munich, 1994, p. 55
[36] H. Rumpf, in: *"Agglomeration"*, W.A. Knepper, Ed., Wiley, New York, 1962, p. 379
[37] D. C.-H. Cheng, *Chem. Eng. Sci.* **1968**, *23*, 1405
[38] D. C.-H. Cheng, *Proc. Soc. Anal. Chem.* **1973**, *10*, 17
[39] K. Kendall, *Powder Metall.* **1988**, *31*, 28
[40] A. Rockenbauer, L. Jókay, B. Pukánszky, F. Tüdős, *Macromolecules* **1985**, *18*, 918
[41] A. Rockenbauer, L. Korecz, B. Pukánszky, *Polym. Bull.* **1994**, *33*, 585
[42] R.K. Mittal, V.B. Gupta, P. Sharma, *J. Mater. Sci.* **1987**, *22*, 1949
[43] D.C. Bassett, *"Principles of Polymer Morphology"*, Cambridge Univ. Press, Cambridge, 1981
[44] H. Schönherr, D. Snétivy, G.J. Vancsó, *Polym. Bull.* **1993**, *30*, 567
[45] B. Pukánszky, *Composites* **1990**, *21*, 255
[46] B. Pukánszky, B. Turcsányi, F. Tüdős, in: *"Interfaces in Polymer, Ceramic, and Metal Matrix Composites"*, H. Ishida, Ed., Elsevier, New York, 1988, p. 467
[47] B. Pukánszky, E. Fekete, F. Tüdős, *Makromol. Chem., Macromol. Symp.* **1989**, *28*, 165
[48] B. Pukánszky, *New Polym. Mater.* **1992**, *3*, 205

Macromol. Symp. **2004**, *214*, 135-145

Morphology and Properties of Poly(oxymethylene) Engineering Plastics

Tatiana Sukhanova,[1] *Vladimir Bershtein,*[2] *Mimi Keating,*[3] *Galina Matveeva,*[1]
Milana Vylegzhanina,[1] *Victor Egorov,*[2] *Nina Peschanskaya,*[2] *Pavel Yakushev,*[2]
Edmund Flexman,[3] *Stefan Greulich,*[3] *Bryan Sauer,*[3] *Kathleen Schodt*[3]

[1] Institute of Macromolecular Compounds of the Russian Academy of Sciences,
31 Bolshoj Pr. V.O, 199004 St. Petersburg, Russia
E-mail: xelmic@imc.macro.ru

[2] Ioffe Physico-Technical Institute of the Russian Academy of Sciences, 26
Politechnicheskaya Str, 194021 St. Petersburg, Russia

[3] E.I. du Pont de Nemours & Company, Experimental Station, Wilmington, DE
19880 323, USA

Summary: Comparative WAXD/SAXS/SEM/DSC structural studies of a series of
semi-crystalline poly(oxymethylene) (POM) engineering plastics, including the
commercial products, homopolymer Delrin® and typical poly(oxymethylene-*co*-
oxyethylene)s, and a few lab-made POM compositions, were performed. The latter
differed in their content of functional additives (present in low concentrations) and
POM molecular weight characteristics. In parallel, their densities, thermal
behavior/laser-interferometric creep rate spectra (DSC/CRS) at 20-180 °C, as well as
long-term creep resistance (LTCR) at 20 °C were studied. It has been found that
introducing the nucleating agents and oxyethylene units resulted in formation of more
fine spherulitic or practically non-spherulitic structure with close- or loose-packed
lamellar stacks. The presence of both "thick" (5-10 nm) and "thin" (1.5-3 nm)
lamellae in the weight ratio of ~3:1 was shown in all cases. Close values of real
POM crystallinities, not exceeding 50%, were obtained by WAXD and DSC. A
predominant role of "straightened out" or slightly bent tie chains in disordered layers
of isotropic POMs was presumed, resulting in segmental dynamics differently
constrained by crystallites (DSC/CRS data). As a result, certain morphology –density
–creep resistance correlations were found.

Keywords: creep; molecular dynamics; morphology; polyethers; thermal properties

Introduction

Poly(oxymethylene) (POM) homo- and copolymers are important engineering plastics. High
POM chain flexibility and symmetry cause their fast crystallization, and complicated morphology
and dynamics at 150-430 K[1-7]. Of most significance for POM engineering properties is the

DOI: 10.1002/masy.200451010

segmental dynamics within disordered regions over the T_g-T_m range, at ~ 300-400 K, which is differently constrained by crystallites. Detailed analysis of dynamics in POMs was recently performed by the combined use of DSC and laser-interferometric creep rate spectroscopy (CRS)[6, 7].

The morphological studies have shown that the spherulitic structure is typical of the melt crystallized POM[8-10]. Using AFM[5], PLM and TEM[8-10], the changes in the crystalline morphology of POM, oriented POM and its blends during crystallization and melting processes have been investigated.

The objective of this study was to understand the influence of the presence of low concentrations of ethylene oxide content in chains or the addition of nucleating agents, and the influence of a decrease in molecular weight on morphology, lamellar textures, thermal properties and creep resistance in molded POM homo- and copolymers. It was intended to reveal structure/properties correlations within a series of commercial products and lab-made compositions.

Experimental

Table 1.   POM samples studied

Sample	M_n	M_w/M_n	Nucleating
Delrin ®IIIP	7.6×10^4	1.84	Nucleated
Delrin 6-1	7.6×10^4	1.84	No nucleating agent
Delrin 6-2	7.6×10^4	1.84	No nucleating agent, minimum amounts of additives
Delrin 6-3	7.6×10^4	1.84	Nucleated
Delrin 6-4	5.5×10^4		No nucleating agent
Delrin 6-5	7.6×10^4	1.84	Nucleated
Celcon M15HP	1.5×10^4	7.47	Nucleated
Celcon M25	8.4×10^3	14.7	Nucleated

Note. Delrin 6-1 – 6-5 samples are the lab-made compositions. Celcons are POMs with 0.4 mol % (Celcon M15HP) or 1.4 mol% (Celcon M25) ethylene oxide units.

A series of POM samples, including the commercial homopolymer Delrin®IIIP, copolymers

Celcon M15HP and Celcon M25, and lab-made POM variations (Table 1), were studied. The latter differed in terms of the functional additives (nucleating agents, process aid, antioxidant) introduced in concentrations less than 1 wt.-%, and by POM molecular weight and MWD. The samples were obtained by injection molding, under pressure of 1 000 bar, with cooling at the rate of 200 K min^{-1}.

Comparative WAXD/SAXS/SEM/DSC structural studies of the materials were performed. In parallel, their densities, thermal behavior/laser interferometric creep rate spectra (DSC/CRS) at 20-180 °C as well as long-term creep resistance (LTCR) at 20 °C, were studied.

The morphology of samples was studied by scanning electron microscopy (SEM). Before SEM observation, samples were etched using the permanganic etching technique[13]. The etched surfaces were sputter coated with a very thin layer of gold (~15 nm thick) to prevent charging during imaging (SCD 050 ion-sputtering device, BALZERS, Switzerland) and were examined with a CamScan MV2300D scanning electron microscope at 15 kV at magnifications from X 500 to X 10 000. X-ray measurements were performed on a DRON-3 diffractometer using Ni-filtered CuK_{α} irradiation. DSC curves were measured using the Perkin-Elmer DSC-2 apparatus, under nitrogen or helium atmosphere, at the heating rates of 2.5-20 K min^{-1}. The CRS method is based on precise measuring of low creep rates as a function of temperature by using a laser interferometer based on Doppler effect[11, 12]. In this research, the CRS setup, operating under uniaxial compression, was used.

Results and Discussion

Morphology

SEM investigations revealed very marked differences in POM morphologies depending on their composition. The majority of the lab-made specimens exhibited well-defined large- or fine-spherulitic morphology (Figure 1 b,c). However, Delrin 6-3 and commercial samples, Delrin®IIIP (Figure 1a) and Celcon M25 (Figure 1d), did not manifest a distinct spherulitic morphology. Delrin 6-3 and Delrin®IIIP consisted of close-packed lamellar stacks. Celcon M25 exhibited the hedritic morphology. On the whole, lamellar stacks in the samples studied had thicknesses ranging from 0.2 to 5 μm. The maximum thickness of 1.8-5.4 μm was found for Delrin 6-3.

Estimation of the average spherulite diameter has shown that introducing the nucleating agents resulted in the formation of fine spherulites ($D_{sph} \approx 20$ μm, Delrin 6-5, Figure 1c) or practically non-spherulitic structure (Delrin 6-3). On the other hand, non-nucleated Delrin 6-2 displays two populations of spherulites with the average diameters of 110 and 55 μm (Figure 1b). Copolymers Celcon M25 and Celcon M15HP distinctly showed more loose-packed lamellar morphologies

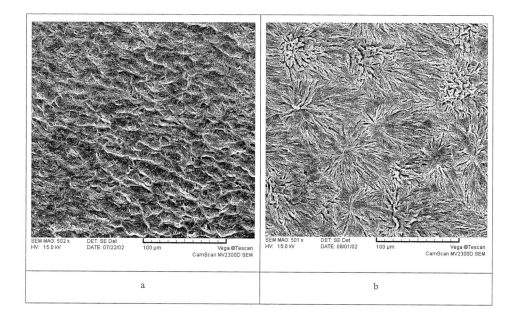

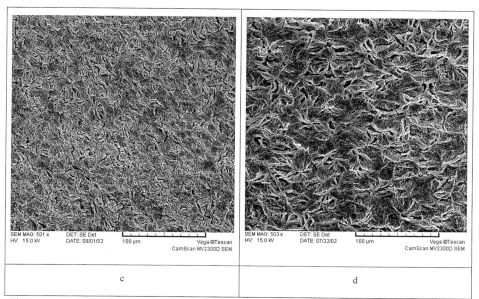

Figure 1. SEM images of the etched cut surfaces of the samples Delrin® IIIP (a), Delrin 6-2 (b), Delrin 6-5 (c) and Celcon M25 (d). Magnification: x500.

than homopolymers (Figure 1d), in accordance with their essentially lower densities (Table 2). On the contrary, Delrins 6-3 and 6-5 had the maximum densities.

Crystalline structure

Structure parameters, estimated by WAXD, SAXS and DSC, and the measured densities are listed in Table 2. In the WAXD profiles of all the samples studied, three typical reflections were observed corresponding to 2Θ = 10.0-10.2°; 23.2° (strong), and 34.5-34.7°. WAXD data were indexed in terms of the most probable hexagonal unit cell with a = 4.42 Å, and c = 17.48 Å.

Table 2.   Structure parameters estimated by WAXD, SAXS, DSC and densitometry data.

Sample	L_c	L_{ab}	X_{WAXD}	X_{DSC}	l_c	L_B	ρ
No.	nm	nm	%	%	nm	nm	g/cm^3
Delrin$^®$IIIP	2.3±0.2	13	46	51±3	9.5±0.5	21.1	1.4197
Delrin 6-1	2.3±0.2	10.5	43	50±3	9.5±0.5	18.9	1.4180
Delrin 6-2	1.8±0.2	13	24-38	49±3	10,0±0,5	18.2	1.4235
Delrin 6-3	1.6±0.2	17.5	46-54	50±3	10.5±0.5	18.2	1.4260
Delrin 6-4	2.2±0.2	15	30-47	49±3	9.5±0.5	19.6	1.4206
Delrin 6-5	1.5±0.2	7	39-50	49±3	10.5±0.5	*	1.4245
Celcon M15HP	2.0±0.2	13	48	48±3	6.5±0.5	17.0	1.4107
Celcon M25	3.0±0.2	20	42	45±3	4.5±0.5	19.6	1.4001

The gyration radii of the scattering domains R_g ranged from 2.8-3.0 nm for Celcons to 3.10 nm (Delrin 6-1 and 6-4), and to 3.33 nm (Delrin$^®$IIIP). The maximum value of the long period $L_B =$ 21.1 nm has also been observed for the commercial Delrin$^®$IIIP. It should be noted that a long period could not be discerned for Delrin 6-5. Combined WAXD/SAXS/DSC analysis indicated the presence of both "thick" (l_c = 5-10 nm, DSC data) and "thin" (L_c = 1.5-3 nm, WAXD data) lamellar crystallites in the POMs studied. Close real values of POM crystallinities (see below) were obtained from both WAXD and DSC estimates. A large spread of X_{WAXD} values was obtained for Delrin 6-2 and Delrin 6-4 only, obviously associated with their large-spherulitic morphology.

Properties

The DSC approach used to analyse POM segmental dynamics and true melting characteristics was similar to that described elsewhere[6, 7]. It included, in particular, the estimation of the effective activation energy Q of segmental motion as a function of temperature, and "parameter of intrachain cooperativity of melting" v, characterizing the length of stereoregular sections in chains[14].

It was found that the endotherm, extending from ~323 to 453 K on the DSC curves of POMs, must be subdivided into three temperature ranges relating to: melting of basic, "thick" lamellae (thickness $l_c \approx$ 5-10 nm, ~433-453 K); melting of "thin" lamellae ($L_c \approx$ 1.5-3 nm, 413-433 K), and the endotherm associated with a gradual "unfreezing" of segmental dynamics within the interlamellar, disordered layers (at ~323-413 K). In all cases, the weight ratio of "thick" and "thin" lamellae was equal to ~3:1.

The found values of the parameter $v \approx$ (2-4)l_c indicated the predominant role of "straightened out" and slightly bent tie chains in intercrystalline layers of isotropic POMs that might imply their constrained dynamics. Really, the "usual" cooperative glass transition (T_g = 250-260 K) was strongly suppressed whereas, depending on the thermal treatment regime, a heat capacity step as a feature of glass transition-like behavior could arise on the DSC curve at any temperature over the ~300-400 K range. By scanning at different scanning rates, the $Q(T)$ dependencies were estimated.

The pronounced $Q(T)$ dispersions were found, indicating large heterogeneity of segmental dynamics over the T_g-T_m range. Thus, Figure 2 shows that Q values range from ~150 to 600 kJ/mol under these conditions. This was interpreted in terms of the variable constraining influence of crystallites on segmental motion in tie chains due to their "anchoring" to the rigid crystalline constraints and some difference in the conformational state of chains. Larger deviations of the activation energies from the Arrhenius $Q(T)$ line corresponded to the stronger constraining effect. Figure 2 shows that the latter is more pronounced in the close-packed Delrin®IIIP than in loose-packed Celcon M25.

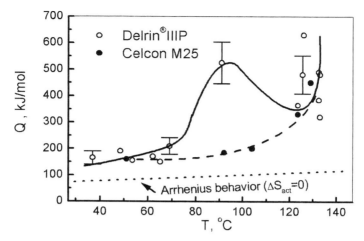

Figure 2. Activation energy of segmental motion versus temperature plots estimated by DSC for Delrin®IIIP and Celcon M25 over the T_g - T_m range. Dotted line corresponds to Arrhenius relation, i.e. to noncooperative relaxation processes, at frequency of 10^{-2} Hz.

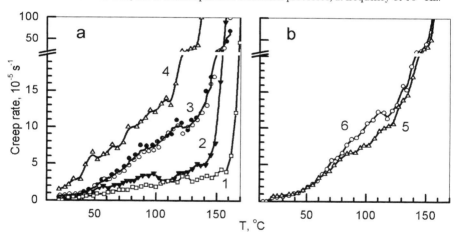

Figure 3. CR spectra obtained for (a) commercial POM plastics, Delrin®IIIP (1, 3) and Celcon M25 (2, 4), and (b) lab-made Delrin compositions with (6-5 sample, curve 5) or without (6-2 sample, curve 6) nucleating agents. Curve 3 indicates reproducibility for two identical samples. Compression, 10 s, 3 MPa (curves 1 and 2) or 10 MPa (curves 3-6).

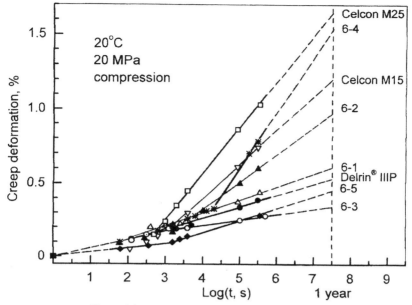

Figure 4. Long-term creep resistance: Comparative POM creep deformation versus logarithm time dependencies.

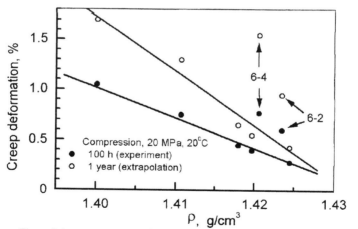

Figure 5. Long-term creep resistance: POM creep deformation versus density dependencies.

The discrete creep rate spectra, with numerous CR peaks in their complicated contours, have previously been obtained under tension for a series of POM compositions[6,7]. These demonstrated both dynamic heterogeneity within the temperature regions of β- and α-relaxations and increasingly constrained segmental relaxations at elevated temperatures. Moreover, it was shown that different small additives affected the CR spectrum.

In this work, the CR spectra, consisting of a few strongly overlapping peaks, were obtained for POMs studied under compression over the 293-443 K range. These spectra also demonsrated a gradual "unfreezing" of differently constrained segmental motions in the interlamellar layers. The spectra of the samples given in Table 1 differed substantially, especially at intermediate temperatures far from both T_g and T_m. Figure 3 illustrates the different relaxation behaviors of commercial Delrin®IIIP and Celcon M25 (a), and the influence of nucleating agent (b). Really, less constraining effect in Celcon than in Delrin (compare with Figure 2) results here in the displacement of the Celcon spectrum by 303-313 K to lower temperatures, i.e., lesser creep resistance at elevated temperatures. Besides, the Celcon spectrum was more sensitive to an increase in stress (Figure 3a). Better behaviour at elevated temperatures was attained in POM due to introducing the nucleating agent, e.g., for the "fine-spherulitic" 6-5 sample compared to that for the non-nucleated, "large-spherulitic" 6-2 sample (Figure 3b).

Finally, the POM compositions studied differed in terms of their long-term creep resistance, even at room temperature, and a certain LTCR versus morphology correlation was found. Figure 4 shows that LTCR increases in the following order. Celcons with loose-packed morphology and large MWD — non-nucleated, "large-spherulitic" Delrins 6-2 and 6-4, with minimum amounts of additives or with the lower molecular weight, respectively, - non-nucleated, "large-spherulitic" Delrin 6-1 - close-packed, nucleated Delrin®IIIP - nucleated Delrin 6-3 and 6-5 compositions with fine or practically non-spherulitic morphologies. Figure 5 shows linear correlation (inverse proportionality) found between LTCR and POM density. Anomalous high creep was observed here only for Delrin 6-4 and Delrin 6-2.

Conclusion

The combined study performed allowed us to determine the differences in morphology of a series of POM (commercial and lab-made) engineering plastics and their correlations with the proprties. Advantages of POM homopolymers compared to the copolymers and the positive role of some small additives in the formation of finer morphology and better creep resistance.

Acknowledgements

This work was supported by Du Pont and CRDF RP1-546-ST-02 project.

[1]   N. G. McCrum, B. E. Read, G. W. Williams, *Anelastic and Dielectric Effects in Polymeric Solids,* Wiley, New York 1967.
[2]   R. J. Hojfors, E. Baer, P. H. Geil, *J. Macromol. Sci., Phys.* **1977**, *13,* 323.
[3]   M. Y. Keating, B. B. Sauer, E. A. Flexman, *J. Macromol. Sci., Phys.* **1997**, *36,* 717.
[4]   B. B. Sauer, P. Avakian, E. A. Flexman, M. Y. Keating, B. B. Hsiao, R. K. Verma, *J. Polym. Sci., Part B: Polym. Phys.* **1997**, *35,* 2121.
[5]   B. B. Sauer, R. S. Mclean, D. J. Londono, *J. Macromol. Sci., Phys.* **2000**, *39,* 519.
[6]   V. A. Bershtein, L. M. Egorova, V. M. Egorov, N. N. Peschanskaya, P. N. Yakushev, M. Y. Keating, E. A. Flexman, R. J. Kassal, K. P. Schodt, *J. Macromol. Sci., Phys.* **2002**, *41,* 797.
[7]   V. A. Bershtein, L. M. Egorova, V. M. Egorov, N. N. Peschanskaya, P. N. Yakushev, M. Y. Keating, E. A. Flexman, R. J. Kassal, K. P. Schodt, *Thermochim. Acta* **2002**, *391,* 227.
[8]   M. Jaffe, B. Wunderlich, *Kolloid. Z.Z. Polym.* **1967**, *216-217,* 203.
[9]   K. O'Leary, P.H. Geil, *J. Macromol. Sci., Phys.* **1967**, *1,* 147.
[10] B. Wunderlich, *Macromolecular Physics, 1-3,* Academic Press, New York 1980.
[11] N. N. Peshanskaya, P. N. Yakushev, A. B. Sinani, V. A. Bershtein, *Thermochim. Acta* **1994**, *238,* 429.
[12] N. N. Peschanskaya, P. N. Yakushev, A. B. Sinani, V. A. Bershtein, *Macromol. Symp.* **1997**, *119,* 79.
[13] R.H. Olley, D.C. Bassett, *Polymer* **1982**, *23,* 1707.
[14] V. A. Bershtein, V. M. Egorov, *Differential Scanning Calorimetry. Physics, Chemistry, Analysis, Technology,* Ellis Horwood, New York 1994.

Linear Viscoelasticity and Non-Linear Elasticity of Block Copolymer Blends Used as Soft Adhesives

Alexandra Roos, Costantino Creton*

Laboratoire de Physico-Chimie Structurale et Macromoléculaire, E.S.P.C.I., 10 Rue Vauquelin, 75231 Paris Cédex 05, France
E-mail : costantino.creton@espci.fr

Summary: The debonding of pressure-sensitive-adhesives from a rigid surface often occurs with the formation of a fibrillar structure bridging the surface and the adhesive. This fibrillar structure is responsible for the large (of the order of 1 kJ/m²) adhesion energy, which is measured, for this type of adhesive despite the very low level of applied stress. One way to account for this dissipated energy is viscous flow in the elongation process of the fibrils. However, for certain types of adhesives, the extension of the fibrils is essentially a quasi reversible elastic process and the energy is only dissipated rather rapidly when the fibrils are detached from the surface. It is therefore essential to characterize the elastic properties of theses adhesives not only in the linear viscoelastic regime but also in the large strain non-linear elastic domain. We have therefore performed tensile tests of model adhesive films based on block copolymers. The respective roles played by the linear viscoelasticity and non-linear elasticity in controlling the properties will be discussed.

Keywords: adhesion; elasticity; fibril; polymer; viscoelasticity

Introduction

When they are removed from a surface, soft adhesives readily form a foamy structure, highly oriented parallel to the tensile direction, which is very analogous to a fibrillar structure[1, 2]. These fibrils have been observed both in peeling experiments[3-5] and in tack experiments[6, 7]. The microscopic processes by which this foam develops have been the focus of many recent studies[8-11] but the actual mechanical properties of the adhesives themselves have only been investigated in the linear viscoelastic regime[12-15]. Yet it is well-known that elongational properties, involving the orientation of the polymer chain, cannot be readily predicted from the linear viscoelastic properties in shear, in particular for polymers with more complicated architectures and very

© 2004 International Union of Pure and Applied Chemistry   DOI: 10.1002/masy.200451011

pronounced elastic effects[16]. One such class of polymers is that of the diblock and triblock copolymers based on styrene and isoprene. For these copolymers, several detailed analyses of their tensile behaviour have been published[17, 18]. However none of theses studies have considered blends of triblock and diblock copolymers and, furthermore, most of these studies have focused on the change in the microphase separated structure which occurs upon deformation[18] and not necessarily on the stress-strain behaviour itself.

We have characterized a family of model pressure-sensitive-adhesives based on block copolymers, not only in the linear viscoelastic regime but also in the large strain regime where non-linear elasticity is dominant. In order to directly relate these results to the adhesive properties, we have performed adhesion experiments in the probe tack geometry as well as a relaxation experiment of the oriented fibrillar foam.

Experimental

The block copolymers were synthesized by anionic polymerization by ExxonMobil Chemical. They are styrene-isoprene-styrene triblock copolymers (SIS) and styrene-isoprene diblock copolymers (SI). Both copolymers contain 15 wt.-% styrene and the molecular weights are 156 kg/mol and 72 kg/mol, respectively. Hence, the diblock is essentially one half of the triblock. In order to obtain PSA properties the block copolymers must be blended with a low molecular weight resin, called a tackifying resin. We used a hydrogenated C-5 resin, commercially available under the trade name Escorez 5380, which was provided by ExxonMobil Chemical. The four model adhesives were prepared by varying the relative amounts of each component. The ratio resin/polymer was kept constant at 60% resin/40% polymer but the diblock/triblock ratio within the polymer phase was varied from 0% diblock to 54% diblock. From now on, the adhesives will be referred to by their diblock content.

All components were dissolved in toluene. The adhesive films (120 μm thick when dry) were obtained by depositing the toluene solution on a standard microscope glass slide and letting the solvent slowly evaporate, first at room temperature and then at 45 °C under vacuum for 24 hours.

Adhesion tests were performed with an instrumented probe tester, where a flat-ended steel cylinder with a polished surface comes in contact with the adhesive film under a controlled pressure and for a given contact time. The probe is then removed from the adhesive at a constant velocity and the force is measured as a function of distance. If the real contact area is known (from a video camera) the force displacement curve can then be transformed into a nominal stress, nominal strain curve, where the strain ε is defined as the displacement of the probe from the time where the force becomes positive (tensile), normalized by the initial thickness of the film. The maximum stress (σ_{max}), maximum strain at detachment (ε_{max}) and adhesion energy W_{adh} are typically extracted from the curves (see Figure 2).

The probe tack tests were performed on the custom made instruments developed in the laboratory[6, 19], with the following experimental parameters: contact pressure: 1 MPa, contact time: 1 s, debonding velocity V_{DEB} varying from 1 to 100 μm/s. The relaxation tests were performed by stopping the driving motor of the probe tack apparatus when the fibrillar foam structure was well formed and monitoring the relaxation of the force under constant displacement conditions.

The tensile tests were performed on a standard tensile tester. The rectangular samples were prepared in a similar way to those for the adhesive tests, but were 300 μm thick and 4 mm wide, and the initial distance between the grips was 15 mm. The crosshead speed was varied between 5 and 500 mm/min, corresponding to an initial strain rate varying between 0.005 and 0.5 s^{-1}. The rheological properties in oscillatory shear were measured on a RDA II Rheometrics parallel plates rheometer at 22 °C. These samples were 2 mm thick and 25 mm in diameter. Care was taken to perform the measurement in the linear regime.

Results and Discussion

Figure 1 shows the modulus G' as a function of frequency for a series of adhesives with different diblock contents. The only adhesive, which clearly appears to have a lower modulus at low frequency, is the 54% diblock. All other adhesives are very similar in G', within that range of frequencies. By contrast, when a probe test was performed at a debonding rate of 100 μm/s, corresponding to an initial strain rate (V_{DEB}/h_0) of 0.8 s^{-1}, the four adhesives could be clearly distinguished, in particular with a different value of maximum tensile stress, σ_{max}, and with a markedly different level of stress in the plateau region of the stress-strain curve. The observed stress level at the plateau and the maximum extension before detachment ε_{max} decrease and increase respectively with increasing diblock content. The video images of the debonding clearly indicate that the plateau stress for these materials corresponds to the formation and elongation of the fibrillar foam. Images show that the average cell size is fairly independent of the diblock content in the blend so that differences in measured stress cannot be due to different microscopic structures of the foam.

Based on these results, it is obvious that the stress level necessary to extend the walls of the foam cannot be controlled by the shear modulus G' only.

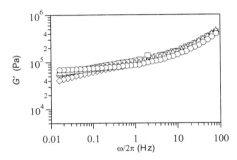 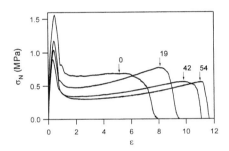

Figure 1. Elastic modulus G as a function of frequency for various SI contents, at T = 22°C. $\bigcirc$: 0 wt.-% SI, $\triangle$: 19 wt.-% SI, $\square$: 42 wt.-% SI, $\diamond$: 54 wt.-% SI.

Figure 2. Nominal stress vs. strain curves at V_{deb} = 100 µm.s^{-1} for the four model blends (stainless steel, RT). Numbers indicate the diblock content of each blend.

In order to understand the results of Figure 2 it is necessary to examine the tensile test results. Figure 3 shows nominal stress vs. strain curves for the same series of adhesives at an initial strain rate of 0.5 s^{-1} (approximately the strain rate applied to the adhesive layer in the tack test at the beginning of the extension process of the foam). Again the four adhesive formulations are very well differentiated and the stress level at large strains decreases with increasing diblock content. An obvious question is, however what the effect of the viscoelasticity is. In that range of frequencies, the small-strain loss modulus G'' of the adhesives is far from being negligible and typically tan δ can be between 0.1 and 1. A good measure of the effect of viscoelasticity can be obtained by performing tensile tests at three different strain rates. Results are shown in Figure 3b. It can be seen that 2 decades in strain rate have less effect than a difference in diblock content. Such an elastic behaviour of the adhesive blend in large strain extension is also confirmed by relaxation tests. If a tack test is stopped while the fibrillar foam is fully formed, the stress does not relax to zero, as one would expect from a liquid capable of flow, but relaxes to approximately 70% of its initial stress before remaining constant. This result shown on Figure 4, clearly demonstrates that the fibrils are able to store elastic energy during their formation and only during detachment or fracture is this elastic energy released.

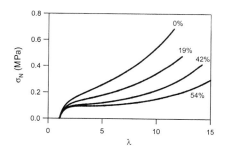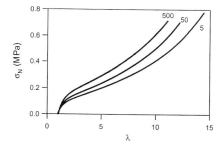

Figure 3a. Tensile curves at a crosshead velocity of 500 mm/min corresponding to an initial strain rate of 0.8 s^{-1} for the four copolymer blends.

Figure 3b. Tensile curves for the 0% diblock blend at crosshead velocities of 5,50 and 500 mm/min corresponding to initial strain rates of 0.008, 0.08 and 0.8 s^{-1}.

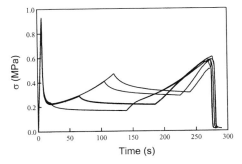
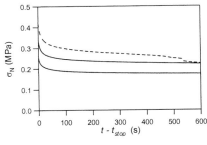

Figure 4a. Stress vs. time curves for probe tests with an intermediate stop for the 0 % diblock adhesive. The test is stopped for 120 seconds, then resumed. Note that the stress-strain curves are almost identical for all tests.

Figure 4b. Stress vs. t-t_{stop} curves obtained by stopping the driving motors in the fibrillation regime at three different stress levels for the 0 % diblock adhesive. The test is stopped here for 600 seconds.

The best way to quantitatively analyze the elastic behaviour in tension of the adhesive blends is with a model describing the non-linear elastic behaviour of rubbers. The microphase separated structure of these adhesives with PS hard domains and a soft PI + resin matrix can be assimilated to a loosely crosslinked rubber where the PS domains would be the crosslink points and the PI chains would provide the entropic springs between them.

The non-linear elastic behaviour of rubbers in uniaxial deformation has of course been the focus of numerous studies in the past[20] and is conveniently analyzed by representing the reduced stress, i.e.

$$\sigma_R = \sigma_N/(\lambda - 1/\lambda^2) \tag{1}$$

where λ is the extension ratio l/l_0, and σ_N is the nominal stress F/A_0. The classical rubber elasticity model, also called the affine model, which considers that all crosslink points are fixed and every segment between crosslink points behaves as an entropic spring, predicts a constant reduced stress with increasing λ[20]. For real rubbery materials this is almost never observed and our adhesives are no exception. In particular, the simple affine model does not take into account the important role-played by entanglements.

More accurate representation of the role played by entanglements in the deformation of rubbers has been the focus of many years of research and has led to the development of several more elaborate models, well reviewed in reference[21]. Among those, the slip-tube model which represents entanglements as links that can slip along the polymer chain but not cross each other and combines the Doi-Edwards tube model with the phantom rubber elasticity model, provides a very clear physical picture of the role of entanglements. Within the framework of this model[21], the reduced stress can be approximately represented by:

$$\sigma_R = G_c + G_e /(0.74\,\lambda + 0.61\lambda^{-1/2} - 0.35), \tag{2}$$

where G_c represents the fixed crosslink points' contribution and G_e represents the entanglements' contribution. When $\lambda \geq 1$, $\sigma_R = G_c + G_e$, so that the sum of both contributions represents the small strain modulus G' measured in oscillatory shear.

A careful observation of equation 2 shows that with increasing λ, the role-played by entanglements decreases and the behaviour becomes dominated by the crosslink points. We have fitted the intermediate (100 – 400 % strain) portion of our stress-strain curves in the simple tensile tests (Figure 3) to the slip-tube model and the results of these fits are shown in figure 5. Clearly, for all adhesive blends, the largest contribution to the small strain modulus G' comes from entanglements and not from crosslink points. Even a substantial change in G_c (at least in relative terms) does not much affect the small strain modulus, in agreement with figure 1. On the other hand, during a tensile test the behaviour appears to be increasingly controlled by G_c as λ increases, which is observed in both Figures 2 and 3.

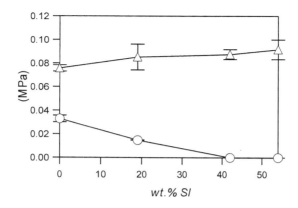

Figure 5.   Slip-tube model coefficients: . —○—: G_c, —△—: G_e, as a function of SI content in the polymer blend.

From a molecular point of view, these results can then be qualitatively interpreted rather simply. The crosslink points represent the network formed by the styrene domains, connected by the isoprene midblocks of the SIS triblock copolymers. When a triblock is replaced by two diblocks, the average density of entanglements (which is controlled by the isoprene/resin structure only) does not change much. On the contrary, the average density of crosslink points decreases sharply when the amount of diblock is increased, reflecting the lesser number of tie molecules between PS domains. If the percentage of diblock is further increased, one expects even the network to disappear altogether when PS domains are no longer connected. Therefore one expects the connectivity between the PS domains (by bridging macromolecular chains) to play a major role in controlling the mechanical and adhesive properties of these soft adhesive blends. Such connectivity is amenable to predictive modelling if the χ parameters between the blend components are known[22].

Conclusions

We have shown that the fibrils which are formed during the debonding of a soft, block copolymer-based PSA from a rigid surface are mainly elastic and can be modelled, as a first approximation, by a constitutive equation describing a network of crosslink points formed by the styrene domains. The behaviour of such a network of PS domains would be analogous to that of a lightly crosslinked rubber network, with a homogeneous molecular weight between crosslinks and a very viscoelastic behaviour at small strains. Since the molecular weight of the isoprene block bridging two styrene domains is much larger than the average molecular weight between entanglements in the isoprene phase, the small strain modulus (in the linear viscoelastic regime) is then essentially controlled by the density of crosslink points, which, in turn, depends on the triblock/diblock ratio in the blend. Therefore, controlling the diblock/triblock ratio provides a convenient way by which to control the non-linear elastic properties of the adhesive, while the resin provides a convenient way by which to adjust the linear viscoelastic properties. Since the two parameters can be varied independently, the optimization of the adhesive properties is easier than for a homopolymer-based system.

Acknowledgements

We gratefully acknowledge funding from the European Commission within the 5th framework programme: Contract N° G5RD-CT 2000-00202. We have also benefited from helpful discussions with Galina Ourieva and Scott Milner from ExxonMobil Chemical and Christophe Derail from the Université de Pau et Pays de l'Adour.

[1]   C. Creton, P. Fabre, in: *"The Mechanics of Adhesion"*, Vol. 1, Elsevier, 2002, pp. 535.
[2]   A. Zosel, *Advances in Pressure Sensitive Adhesive Technology* **1992**, *1*, 92.
[3]   L. Benyahia, C. Verdier, J. M. Piau, *J. Adhes.* **1997**, *62*, 45.
[4]   S. F. Christensen, H. Everland, O. Hassager, K. Almdal, *Int. J. Adhesion & Adhesives* **1998**, *18*, 131.
[5]   C. Derail, A. Allal, G. Marin, P. Tordjeman, *J. Adhes.* **1997**, *61*, 123.
[6]   H. Lakrout, P. Sergot, C. Creton, *J. Adhes.* **1999**, *69*, 307.
[7]   A. Zosel, *J. Adhes.* **1989**, *30*, 135.
[8]   K. Brown, J. C. Hooker, C. Creton, *Macromolecular Materials and Engineering* **2002**, *287*, 163.
[9]   C. Creton, H. Lakrout, *J. Polym. Sci., Part B: Polym. Phys.* **2000**, *38*, 965.
[10]   A. Chiche, P. Pareige, C. Creton, *Comptes Rendus de l'Académie des Sciences de Paris, Série IV* **2000**, *1*, 1197.

156

[11]   I. Chikina, C. Gay, *Phys. Rev. Lett.* **2000**, *85*, 4546.

[12]   D. H. Kaelble, *J. Macromol. Sci., Rev. Macromol. Chem.* **1971**, *C6*, 85.

[13]   M. Sherriff, R. W. Knibbs, P. G. Langley, *J. Appl. Polym. Sci.* **1973**, *17*, 3423.

[14]   G. Kraus, K. W. Rollmann, *J. Appl. Polym. Sci.* **1977**, *21*, 3311.

[15]   D. W. Aubrey, M. Sherriff, *J. Polym. Sci., Part A: Polym. Chem.* **1978**, *16*, 2631.

[16]   S. F. Christensen, G. H. McKinley, *Int. J. Adhesion & Adhesives* **1998**, *18*, 333.

[17]   C. C. Honeker, E. L. Thomas, *Chemistry of Materials* **1996**, *8*, 1702.

[18]   E. Prasman, E. L. Thomas, *J. Polym. Sci., Part B: Polym. Phys.* **1998**, *36*, 1625.

[19]   G. Josse, P. Sergot, M. Dorget, C. Creton, submitted to Journal of Adhesion.

[20]   L. R. G. Treloar, in: *"The physics of rubber elasticity"*, Clarendon Press, Oxford 1975, pp. 210.

[21]   M. Rubinstein, S. Panyukov, *Macromolecules* **2002**, *35*, 6670.

[22]   K. Daoulas, D. N. Theodorou, A. Roos, C. Creton, submitted to Macromolecules.

Macromol. Symp. **2004**, *214*, 157-171

Micromechanical Mechanisms for Toughness Enhancement in β-Modified Polypropylene

Sven Henning,[1] *Rameshwar Adhikari,*[1] *Georg H. Michler,*[*1]
Francisco J. Baltá Calleja,[2] *József Karger-Kocsis*[3]

[1]Institute of Materials Science, Martin Luther University Halle Wittenberg, D-06099 Halle, Germany
E-mail: michler@iw.uni-halle.de
[2]Instituto de Estructura de la Materia, CSIC, Serrano 119, 28006 Madrid, Spain
[3]Institute of Composite Materials, University of Kaiserslautern, D-67633 Kaiserslautern, Germany

Summary: Two samples of isotactic polypropylene (iPP) representing the crystalline α- and β- modifications are compared with regard to their semicrystalline morphology and the resulting micromechanical mechanisms. Processes at room temperature (23°C) and at −40°C have been investigated by means of different microscopic techniques. Good results can be achieved by the application of a chemical etching technique to the deformed samples prior to scanning electron microscopy inspection. The toughness enhancement that is measured for the β-iPP is attributed to micromechanical mechanisms initiated within the intercrystalline, amorphous phase. It is shown that the specific nanostructure (lamellar arrangement) causes significant changes in the mechanical behaviour of the materials.

Keywords: deformation behaviour; electron microscopy; microhardness; nanostructure; polypropylene; semicrystalline polymers

Introduction

Isotactic polypropylene (iPP) belongs to the family of commodity plastics and occupies the third place after low-density polyethylene (LDPE) and polyvinyl chloride (PVC) in the worldwide plastic market. The most important applications of this polymer are the production of films, textile fibres and moulded parts in which the plastic is processed mainly by extrusion and injection moulding.[1, 2] The pronounced diversity of this polymeric material is also connected with the existence of several crystal modifications.[3] Of particular interest from a technical point of view is the β-modification of PP because of its enhanced toughness compared to the ordinary α-modification.[4, 5] The copolymers, blends and composites containing iPP find application in several fields of everyday life. The aim of this work is to analyze the micromechanical mechanisms responsible for the enhanced toughness of the β-iPP compared to that of common α-modification.

DOI: 10.1002/masy.200451012

Experimental

Materials

Polypropylene samples were prepared from two high molecular weight commercial isotactic polypropylene grades (BOREALIS AG): α-iPP was Daplen BE50 and β-iPP was Daplen BE60. Sheets of a thickness of 1 mm were produced using a hot press and subsequent multistage crystallization procedure in the second β-iPP case. Miniaturized, dumbbell-shaped tensile bars were punched out of the plates using a special pierce tool.

Mechanical Testing

A MINIMAT miniature materials testing device (POLYMER LABORATORIES, UK) was used to perform tensile deformation experiments at a cross-head speed of 1 mm/min at room temperature. Stress-strain diagrams were recorded to qualitatively follow the states of deformation. They will not be discussed in detail in this work because they represent single events and the experiments do not follow any standard specification. A custom-made heating/cooling chamber adapted for the MINIMAT device was used to perform tensile experiments at –40°C. The cooling was provided by a nitrogen flow. The temperature was measured close to the sample surface by means of a digital thermocouple thermometer. Samples were elongated until fracture without recording the stress-strain-curves.

Scanning Electron Microscopy (SEM)

In order to obtain a planar geometry and flat surfaces, the deformed samples were embedded in epoxy resin and microtomed down to the middle section of the tensile bar using a LEICA microtome with a metal blade. Regions of interest corresponding to the different states of deformation were selected for SEM investigations (Fig. 1). Subsequently, a modified permanganic etching technique according to OLLEY and BASSETT [6] was applied to reveal the morphology of the deformed material. For our procedure, a mixture of 5 ml distilled water and 50 ml concentrated sulphuric acid plus 0.55 g $KMnO_4$ was adequate. We found that an etching time of 15 to 30 minutes gave optimum results. To avoid electrical charging, samples were sputter-coated with a gold layer of 12 nm (EDWARDS sputter coater). Micrographs of the deformation structures were recorded using a JEOL JSM 6300 scanning electron microscope operated at 15 kV accelerating voltage.

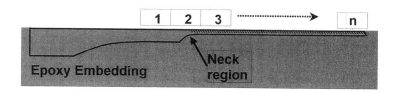

Figure 1.   Outline of the preparation procedure. The embedded tensile bar is microtomed down to the middle.   After chemical etching, SEM micrographs were taken at regions representing different states of deformation (1, 2, 3, ... n).

Scanning Force Microscopy (SFM)

The morphology of untreated surfaces of α- and β-iPP samples crystallized from the melt were imaged using a DIGITAL INSTRUMENTS SFM equipped with a NANOSCOPE IIIa controller. A silicon cantilever with a resonance frequency of approx. 300 kHz and a spring constant of approx. 15 N/m was used. Best results concerning the information on the semicrystalline morphology were obtained in the tapping mode. Only phase images are presented.

Transmission Electron Microscopy (TEM)

From non-deformed and from elongated samples similar to those used for SEM investigations, ultrathin sections were prepared from regions of interest close to the middle of the tensile bar. The materials were stained in RuO_4 vapour for several hours. For the preparation of the approx. 100 nm thick ultrathin sections, a LEICA ULTRACUT microtome equipped with a DIATOME diamond knife was used. Micrographs were acquired using a JEOL JSM 2010 microscope operated at 200 kV accelerating voltage.

Microhardness

In order to obtain a co-planar geometry and flat surfaces, the deformed samples were embedded in epoxy resin and microtomed down to the middle section of the tensile bar using a LEICA microtome equipped with a metal blade. Surface effects (skin morphology or scratches) were thereby eliminated. For the microindentation experiments the regions corresponding to the different states of deformation were selected (similar to SEM preparation, Fig. 1).

Microhardness measurements were performed at room temperature using a LEITZ microhardness tester equipped with a Vickers square pyramidal diamond indenter (168°). Indentation experiments were performed using a load of 98.1 mN and a loading time of 0.1 min. The final permanent deformation was measured immediately after load release using the micrometer eyepiece of the microscope. Microindentations were performed starting from undeformed regions and proceeding in equidistant steps of 100 to 150 µm across the neck region into the cold drawn area. The hardness values expressed in [MPa] were calculated from the residual projected area of the indentation using the relation:

$$H = kP/d^2 \tag{1}$$

where d is the length [µm] of the impression diagonal, P is the load [N] applied and k is a geometrical factor equal to 1.854×10^6.[7] The hardness values were derived from the indention diagonals parallel $H_{\parallel}$ and normal $H_{\perp}$ to the straining direction. Each data point represents the average of at least four indentations.

Morphology

The most significant morphological feature of the more common α-form of the iPP is the so-called "cross-hatched" arrangement of the crystalline lamellae, as shown in Fig. 2. The main lamellae ("mother lamellae") grow radically from an initial site (centre of the spherulite), whereas the "daughter lamellae" are formed by an epitaxial growth onto them, exhibiting a typical angle of 81°[8]. In contrast to that phenomenon, a stacked, parallel arrangement of bundles of lamellae is found for the β-modified material (Figs. 2, 3). Here, the lamellae form a more sheaves-like superstructure rather than a pure radiating, spherulitic growth. From the SFM and TEM micrographs it becomes clear that the intercrystalline, amorphous portion is distributed more continuously in case of the β-iPP. In other words, the latter shows the typical lamellar structure as it is at hand in other major semicrystalline polymers (LDPE, HDPE, UHMWPE).[9, 10]

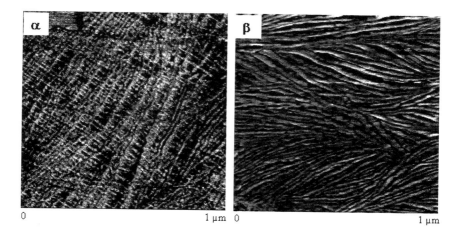

Figure 2.   Comparison of the semicrystalline morphologies of α - and β- iPP. The arrangements of the crystalline lamellae (brighter shades of grey) and the amorphous portion (dark) differ significantly. Tapping mode SFM images, phase contrast.

The crystalline lamellae of the β-iPP are slightly thicker (approx. 15 nm) than the lamellae found in the α-iPP (approx. 10 nm). Due to the nucleating agent that was added by the producer in the case of the β-modified material, the size of the spherulites is different for the two materials. For the α-iPP the average diameter is approx. 50 μm, for the β-iPP the average diameter is reduced to approx. 10 μm. From the SEM micrographs we conclude that the β-modified material is "pure", since the whole area of the sample is filled with spherulitic superstructures that can be assigned to the crystalline β-modification.

Figure 3. Stack-like arrangement of crystalline lamellae (bright) and intercrystalline, amorphous material (dark) in the β-modification of polypropylene. TEM micrograph, RuO₄ stained.

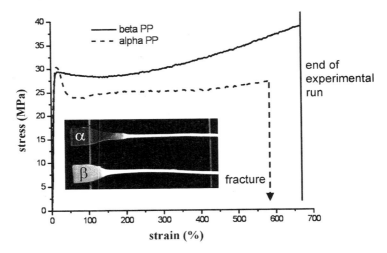

Figure 4. Comparison of the stress-strain curves obtained for miniaturized tensile bars at room temperature. Due to the geometrical limitation of the tensile device the β-iPP bar could not be drawn up to fracture.

Deformation at 23°C: Comparison of the Micromechanical Mechanisms

The plastic deformation of the tensile bars of both PP types under load proceeds via formation and propagation of two necks in opposite directions. The tensile stresses at yield are comparable (approx. 30 MPa in both cases), and the curves for the elastic part before the yield point coincide. Due to the geometrical limitations of the miniature tensile tester, the β-iPP sample was not drawn up to the fracture. The strain at break for the α-iPP sample is approx. 600 %. If we consider that the area below the stress-strain curves can be taken as a measure for the energy that is dissipated by plastic deformation within the sample, then the β-iPP is superior.

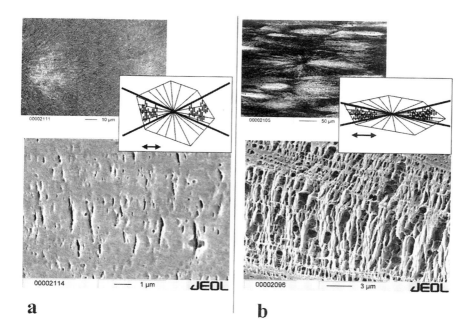

a   **b**

Figure 5.   Micromechanical processes in α-iPP for initial stages of deformation at 23°C. The approximate elongations represented in the micrographs are $\lambda \cong 1.1$ (a) and $\lambda \cong 2$ (b), respectively. The lower images are higher magnifications from selected areas in the overview images. SEM micrographs after chemical etching.

In Fig. 5 the structural changes within the sample are shown for initial stages of deformation, i.e. for elongations up to $\lambda \cong 2$. The formation of microvoids is limited to polar regions of the spherulites (with respect to the straining direction). At higher magnifications the craze-like nature of the deformation structures becomes visible. The formation of microvoids is coupled with the fibrillation of the material. At an elongation of $\lambda \cong 2$ the mesh-like structure of the crosshatches lamellar arrangement is still to be seen (Fig. 5b).

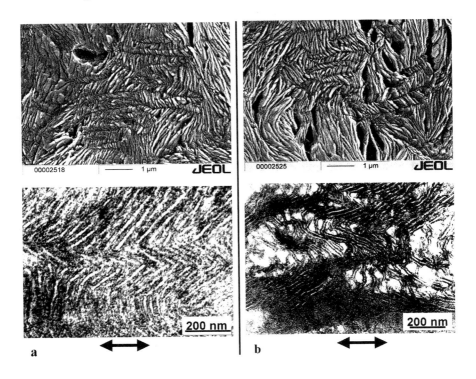

Figure 6.   Micromechanical processes in β-iPP for initial stages of deformation at 23°C. The approximate elongations represented in the micrographs are $\lambda \cong 1.2$ (a) and $\lambda \cong 1.4$ (b), respectively. The upper images are SEM micrographs after chemical etching and the lower ones are TEM results obtained for similar regions (RuO_4 stained).

In contrast to the results for the α-iPP, the whole sample area contributes to micromechanical processes that are connected with microvoid formation and fibrillation. We have found no dependency of the intensity of microvoiding from the actual situation of the investigated area with respect to the straining direction. In Fig. 6 the structures observed at elongations of $\lambda \cong$

1.2(a) and $\lambda \cong 1.4$ (b), respectively, are given. In this case the orientation of the lamellae was perpendicular to the straining direction. Two general processes are observed concurrently: the formation of a so-called "chevron"-pattern and lamellar separation. The first process can be interpreted as a collective twisting of bundles of lamellae. There is neither microvoid formation nor fibrillation. The most interesting aspect is that similar processes are observed for other materials with lamellar morphologies, as there are other semicrystalline polymers (such as polyethylenes [9, 10]) and styrene/butadiene block copolymers[11]. The second process includes the formation of microvoids and fibrillation. Lamellar separation is dominated by the deformation of the intercrystalline, amorphous material (Fig. 9). The actual nanostructure, i.e., the lamellar arrangement of a hard (crystalline) and a soft (amorphous) phase connected by entangling as well as tie molecules dominates the deformation process. Structural features in the range of 10 nm initiate craze-like deformation processes. These crazes are highly localized zones of plastic deformation, but the small size and the large number make that a great part of the material contributes to deformation.

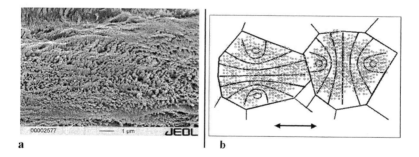

a   b

Figure 7.   Micromechanical processes in β-iPP for initial stages of deformation at 23°C (REM, chemical etching). The approximate elongation λ represented in the micrograph is approx. 2 (a). The sketch in (b) indicates that, independent of the orientation of the lamellae with respect to the straining direction, intensive microvoid formation takes place in the whole volume of the material.

In Fig. 7 the crystalline lamellae are aligned parallel to the straining direction. It is observed that the lamellae are broken in fragments. Once more, intensive microvoid formation takes place. The different micromechanical mechanisms that are observed for initial stages of deformation at 23°C are outlined in Fig. 8.

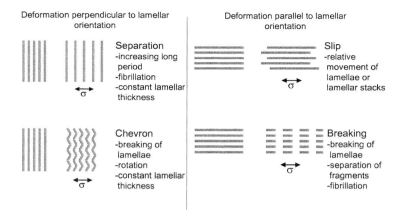

Figure 8. Comparison of different micromechanical processes induced by the lamellar morphology of the β-iPP at initial stages of deformation depending on the loading direction relative to the lamellar orientation. To some extent they are competing processes (chevron formation and lamellar separation), to some extent successive (slip before breaking).

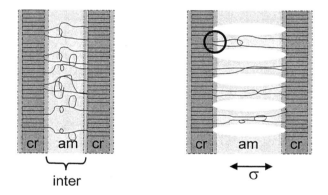

Figure 9. Illustration of the process of lamellar separation. The intercrystalline, amorphous portion is fibrillated, and nanoscopic voids are formed. For the initial deformation stage depicted here the crystalline lamellae seem to stay intact. The situation can be interpreted as a craze that is initiated by the actual semicrystalline nanostructure.

Microhardness

Starting from initial hardness values of 110 MPa for the α-iPP and 100 MPa for the β-iPP there is a significant drop as the stress whitening region or neck region is crossed. At elongations of $\lambda \cong 2$ there is a minimum in the microhardness. From the comparison of SEM images representing similar states of deformation we conclude that the drop in the microhardness is mainly caused by the intensive microvoid formation that is observed. As the deformation proceeds, an indentation anisotropy occurs, and the hardness values increase is phenomenon can be attributed to the onset of strain hardening and orientation of the material, and will be discussed elsewhere.[12]

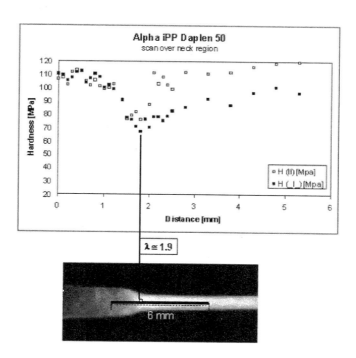

Figure 10. Changes of the microhardness (H) of α-iPP with the progression of deformation (23°C). The H values decreases significantly from an initial value of approx. 110 MPa to a minimum of approx. 70 MPa as the neck region (beginning of stress whitening) is crossed.

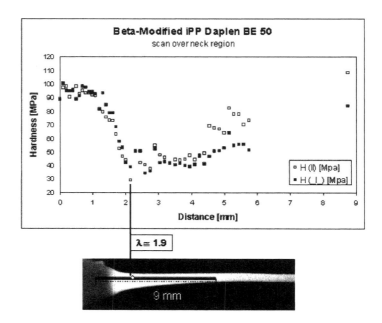

Figure 11. Changes of the Microhardness (H) of β-iPP with the progression of deformation (23°C). The magnitude of H decreases dramatically from an initial value of approx. 100 MPa to a minimum of approx. 35 MPa on going through the neck region (beginning of stress whitening).

Although the hardness curves for both the materials examined are similar, the drop in hardness that is observed for the β-modification is much greater (Fig. 11). This observation correlates well with the micromechanical processes connected with intensive microvoid formation for the β-iPP. This is in good agreement with the SEM results, showing that more sample volume contributes to microvoiding in the case of the β-iPP.

Deformation at -40°C: Comparison of the Micromechanical Mechanisms

At temperatures below the glass transition temperature (T_g) of the amorphous phase (approx. 0°C) polypropylene tends to become brittle. Toughness enhancement measures also have overcome the problems related to low temperature applications.

For both of the systems investigated, the material becomes brittle when tested at –40°C. Typical crazes are observed. The crazes are not influenced by the spherulitic morphology; they run perpendicular to the straining direction, and the edges are bridged by fibrils (Fig. 12). The crazes in the β-modified PP are finer and shorter, and there are much more to be seen in a similar sample area. That means that even at low temperatures more plastic deformation can take place as crazing is the major energy dissipating process. The multiple crazing should result in a tougher material. Nevertheless, none of the nanostructure-controlled processes discussed earlier were found in the samples deformed at –40°C. These nanomechanical processes are initiated by the more mobile amorphous material, and they are freezing as the loading temperature is below the T_g of the amorphous phase.

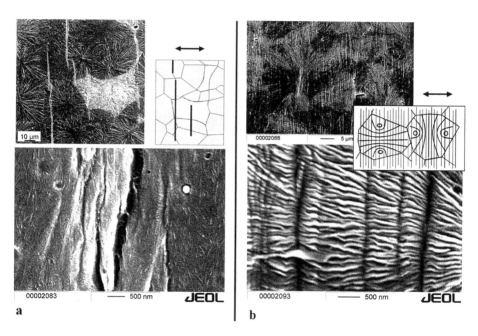

Figure 12. Comparison of the micromechanical mechanisms observed at –40°C. Only few, but long and coarse, crazes are typical for the α-modification (a). Fibrillation is clearly to be seen (a, lower image). The crazes generated in the β-iPP are smaller and thinner (b). Due to a large number of crazes, more material contributes to plastic deformation. SEM after chemical etching.

Conclusions

Although there is a number of publications on the deformation behaviour of β-iPP[13-20] the direct comparison of the two types of polypropylene that is presented here leads to new aspects to clarify the reasons for the different mechanical properties.

Room Temperature (23°C)

For initial stages of deformation ($\lambda \leq 2$), micromechanical processes initiated by the intercrystalline, amorphous material are dominant. The enhanced toughness of the β-iPP can be explained to a great extent by the general differences in the micromechanical mechanisms that are observed by SEM and TEM. The special arrangement of the crystalline lamellae, i.e., the lamellar nanostructure of the β-modified PP, activates processes of plastic deformation that are highly locallized on the nanoscale, but huge in number. They are homogeneously distributed within the volume of the stressed sample. The general processes are:

-lamellar separation with craze-like microvoid formation and fibrillation

-chevron formation

-lamellar slip

-breaking of lamellae with microvoid formation.

Sub-T_g Temperature (-40°C)

Below the T_g of the amorphous phase the intensive micromechanical processes that are responsible for the superior mechanical properties of β-iPP are freezing. Instead of the nanostructure-controlled micromechanisms, ordinary crazing takes place that can be attributed to a brittle behaviour. Nevertheless, there are many more crazes detected within the β-modification, which are thinner and shorter than that in the α-iPP. More investigations are in progress to clarify the role of this multiple crazing effect in β-iPP for toughness enhancement at low temperatures.

[1]   H.G. Karian (Ed.), *"Handbook of polypropylene and polypropylene composites"*, Marcel Dekker, New York 1999.
[2]   J. Karger-Kocsis (Ed.), *"Polypropylene: structure, blends and composites"*, Vol. 1. Structure and morphology, Chapman & Hall, London 1995.
[3]   A. Turner Jones, J. M. Aizlewood, D. R. Beckett, *Makromol. Chem.* **1964**, *75*, 134.
[4]   C. Grein, C. J. G. Plummer, H. H. Kausch, Y. Germain, P. Béguelin, *Polymer* **2002**, *43*, 3279.
[5]   H. B. Chen, J. Karger-Kocsis, J. S. Wu, J. Varga, *Polymer* **2002**, *43*, 6505.
[6]   R. Olley, D. C. Bassett, P. J. Hine, I. M. Ward, *J. Mat. Sci.* **1993**, *28*, 1107.
[7]   F. J. Baltá Calleja, S. Fakirov, *"Microhardness of Polymers"*, Cambridge University Press, Cambridge 2000.
[8]   R. A. Phillips, M. D. Wolkowitcz, in: *"Polypropylene Handbook"*, Edward P. Moore, Jr., Ed., Carl Hanser Verlag, Munich 1996, p. 113ff.
[9]   G. H. Michler, K. Morawietz, *Acta Polymerica* **1991**, *42*, 620.
[10]   G. H. Michler, *"Kunststoff-Mikromechanik: Morphologie, Deformations- und Bruchmechanismen von Polymerwerkstoffen"*, Carl Hanser Verlag München 1992, p. 212ff.
[11]   R. Adhikari, G. H. Michler, T. A. Huy, E. Ivan´kova, R. Godehardt, W. Lebek, K. Knoll, *Macromol. Chem. Phys.* **2003**, *204*, 488.
[12]   S. Henning, G. H. Michler, F. Ania, F. Baltá Calleja, *J. Coll. Polym. Sci*, submitted.
[13]   P. T. S. Dijkstra, D. J. van Dijk, J. Huétink, *Polym. Eng. Sci.* **2002**, *42*, 152.
[14]   M. Aboulfaraj, C. G´Sell, B. Ulrich, A. Dahoun, *Polymer* **1995**, *36*, 731.
[15]   J. X. Li, W. L. Cheung, *Polymer* **1998**, *39*, 6935.
[16]   J. X. Li, W. L. Cheung, C. M. Chan, *Polymer* **1999**, *40*, 2089.
[17]   J. X. Li, W. L. Cheung, C. M. Chan, *Polymer* **1999**, *40*, 3641.
[18]   F. Chu, T. Yamaoka, H. Ide, Y. Kimura, *Polymer* **1994**, *35*, 3442.
[19]   F. Chu, T. Yamaoka, Y. Kimura, *Polymer* **1995**, *36*, 2523.
[20]   T. Yoshida, Y. Fujiwara, T. Asano, *Polymer* **1983**, *24*, 925.

Macromol. Symp. **2004**, *214*, 173-196   173

Morphology and Micromechanical Behaviour of SBS Block Copolymer Systems

R. Adhikari,[1] *R. Godehardt,*[1] *W. Lebek,*[1] *S. Goerlitz,*[1] *G. H. Michler,**[1] *K. Knoll*[2]

[1] Institute of Materials Science, Martin-Luther University Halle-Wittenberg, D-06099 Halle/Saale, Germany

[2] BASF Aktiengesellschaft, Polymer Laboratory ZKT/I-B1, D-67056 Ludwigshafen, Germany

Summary: By means of transmission electron microscopy (TEM) and uniaxial tensile testing, the connection between the morphology and the micromechanical properties of selected styrene-rich styrene/butadiene block copolymers was studied with respect to their molecular architecture. In particular, the structure-property correlation of a lamellae forming asymmetric linear SBS triblock copolymer was examined by systematically varying the sample preparation techniques and testing temperature. The molecular architecture was found to influence directly the morphology formation of the block copolymers. Different mechanisms such as drawing of the lamellae, shearing in the rubbery phase and rotation of the lamellar axis were observed. From room temperature down to the temperature close to glass transition temperature of the soft phase, a homogeneous plastic drawing of glassy lamellae was perceptible.

Keywords: blends; electron microscopy; mechanical properties; micromechanisms; morphology; SBS block copolymers

Introduction

Block copolymers lie at focus of intensive research activities in the contemporary macromolecular science and technology. This is attributable to a wide range of fascinating fundamental issues associated with the understanding of self-assembly processes, their potential application possibilities in nanotechnology etc. The concerns regarding the past successes and future perspectives as well as the challenges in the field of these nanostructured heteropolymers have been addressed recently by Lodge [1].

At sufficiently high molecular weight and low polydispersity, the block copolymer molecules are known to self-assemble into a variety of ordered structures in the melt and solid state by a process called microphase separation transition. The latter is a consequence of the intramolecular phase separation between homopolymer chains linked together by means of a covalent bond [2].

    DOI: 10.1002/masy.200451013

Mechanical properties, especially the balance of strength, stiffness and toughness, are important for many everyday applications of polymeric materials. The styrene/butadiene block copolymers, which assure the transparency of the products due to the presence of inherent nano-scale heterogeneity, offer the possibility of tailoring application relevant mechanical properties by suitably choosing the molecular and processing parameters. An inevitable aspect in the structure-properties correlations of the block copolymers is the understanding of micromechanical processes of deformation, which firmly bridges the mechanical properties with polymer morphology.

Styrene/diene block copolymers (dienes: butadiene or isoprene) are seldom used as pure materials [3,4]. Instead, they are alloyed with other polymers such as polystyrene homopolymer. The objective of mixing these copolymers into standard polystyrene is to toughen the latter while preserving the transparency. Due to higher production costs of block copolymers, it is desired to enhance the toughening ability of block copolymers, i.e., to develop the copolymers which are better compatible with polystyrene so that less amounts of block copolymer is needed to achieve a good level of stiffness/toughness ratio. In addition, one is interested to keep the diene content as low as possible to suppress the ease of thermal degradation of the products without making any compromise in the toughness level. To meet these demands, in recent years, new types of styrene/butadiene block copolymers have been developed [4-6] and even introduced as commercial products in the market. Special emphases have been put on designing copolymers having variable molecular architectures, e.g., asymmetric conformation, modified interfacial profile etc.

The aim of this paper is to discuss how the molecular architecture and the resulting morphology of selected polystyrene-b-polybutadiene-b-polystyrene (SBS) triblock copolymers are connected with their mechanical and micromechanical behaviour. A brief review of the structure-property correlation in styrene/diene block copolymers is given and the results on styrene-rich block copolymers (total styrene content, $\Phi_{PS} \sim 70\%$) are presented in detail.

Structure-Property Correlation in Block Copolymers

Generally, nanostructures that are formed in the solid state in block copolymers are of practical interest. In the classical sense, these structures are adjusted by changing the relative composition of the chemically linked chains. Figure 1 shows schematically the morphologies observed in classical two-component block copolymers (e.g., both styrene/butadiene di- and triblocks). These morphologies in styrene/diene block copolymers are responsible for different kinds of micromechanical mechanisms and mechanical behaviour. This fact illustrates the significance of morphology control in block copolymers from practical point of view. It should be stressed that, for good mechanical properties of styrene/butadiene block copolymers, the rubbery polybutadiene (PB) block should be anchored on both the ends by polystyrene (PS) blocks that are glassy at room temperature.

bcc spheres   *hex cylinders*   *lamellae*   *hex cylinder*   *bcc spheres*

Figure 1. — Scheme showing the classical block copolymer morphologies with decreasing □ volume content, where □ and ■ represent different constituents.

In diblock copolymers, the nature and the dimension of the microphase separated structures are adjusted by changing composition and molecular weight of the constituents. In the strong segregation limit, with increasing polystyrene (PS) content, body centred cubic spheres, hexagonal arranged cylinders and three dimensional 'gyroid' network of polystyrene (PS) domains dispersed in the matrix of polyisoprene were observed in polystyrene-b-polyisoprene (SI) diblock copolymers [9]. With further increase in the PS content, the alternating layers of PS and PB lamellae, and then the structures mentioned above in the reversed order were found.

Further possibilities for the fine-tuning of block copolymer nanostructures are provided by blending and processing [3-7]. Three-component block copolymers and their derivatives form an additional variety of microphases, which may allow novel routes to developing materials having new property profile [7-9]. It has been known for some time that the modification of molecular architecture may significantly alter their phase behaviour and opens up new horizon of morphology control in these copolymers [6,10-12]. Due to architectural modifications of block copolymers, limitations because of composition (e.g., in an AB diblock) can be overcome to achieve a particular morphology and, hence, improved mechanical properties. These new morphologies and new routes to control block copolymer nanostructures have posed new challenges for polymer scientists to clarify their complex structure-property correlations.

Experimental

Materials

The important characteristics of the block copolymers studied are collected in Table 1. The details of the morphology of respective copolymers may be found in the references mentioned in the parentheses. The samples provided by the BASF Aktiengesellschaft, Ludwigshafen were synthesised by butyl-lithium initiated living anionic polymerisation. Asymmetric star block copolymers (ST1-S74 and ST3-S74) were prepared by joining the living S_1BS_2 or $S_1(S/B)S_2$ chains (where S/B stands for a polystyrene-co-polybutadiene random copolymer; the polystyrene blocks S_1 and S_2 are of different lengths) using oligofunctional coupling agent. Except LN1-S74, which is a linear symmetric SBS triblock copolymer, the remaining block copolymers possess non-classical molecular structure (asymmetric structure, tapered transition, etc., see table 1). Note that the net composition of the copolymers studied is nearly constant (ca. 70 vol.-% styrene).

One of the copolymers chosen to prepare blends with homopolystyrene hPS (PS015, PS033 and PS190) by solution casting was an asymmetric star block copolymer (ST3-S74, volume fraction of PS, Φ_{ST} = 0.74; see Table 1). Each star block copolymer has 4 asymmetric SBS arms in average, one of which is much longer than the others. The longest arm is styrene-rich while the shorter ones are butadiene-rich [6]. Three different types of hPS samples were used: PS015 (M_w =

15 000 g/mol and M_w/M_n = 1.29), PS033 (M_w = 33 000 g/mol and M_w/M_n = 1.81) and PS190 (M_w = 190 000 g/mol and M_w/M_n =2.3). The first two have molecular weights much smaller than that of the longest PS outer block of the star block copolymer while the third one has an M_w higher than that of all the PS blocks.

Sample Preparation

Table 1.   The characteristics of the block copolymers studied; LN and ST stand for linear and star architectures, respectively.

Samples	[§]M_w (g/mol)	[§]M_W/M_n	*Φ_{PS}	Remarks
LN1-S74	87 700	1.07	0.74	symmetric SBS triblock, sharp transition with the pure PB mid-block, linear architecture [12]
LN2-S74	105 100	1.13	0.74	asymmetric S_1BS_2 triblock copolymer, ($S_1 \neq S_2$), tapered transition [12,13]
LN4-S65	139 200	1.20	0.65	random PS-co-PB copolymer as rubbery mid-block, symmetric outer PS blocks, linear architecture [5,13]
ST1-S74	9 800	1.99	0.74	highly asymmetric star architecture, each arm with S_1BS_2 structure ($S_1 \neq S_2$), sharp transition
ST3-S74	100 000	1.91	0.74	Structure similar to ST1, each arm with $S_1(S/B)S_2$ structure ($S_1 \neq S_2$), S/B is a random copolymer of PS and PB [6]

[§]   number average (M_n) and weight average (M_w) molecular weight determined by gel permeation chromatography using PS calibration

•   total styrene volume content determined by Wijs double bond titration procedure

The samples were prepared by three different methods: solution casting (using toluene as solvent), compression moulding (melt casting at 190°C and 200 bar pressure) and injection moulding (mass temperature 250°C and mould temperature 45°C). Solution cast films were prepared form 3% (weight/volume) solution of each polymer in toluene allowing the solvent to evaporate over a period of about a week followed by vacuum drying at 23°C for several days and finally annealing for 48 hours in a vacuum oven at 120°C.

Tensile Testing

Macroscopic tensile tests using injection and compression moulded bars and solution cast films were carried out using universal tensile machine (Zwick 1425 or Instron 1407) at a cross head speed of 50 mm/min at different temperatures. At least 6 specimens of each sample were tested.

Transmission Electron Microscopy (TEM)

Morphological details of the samples were investigated by means of a transmission electron microscope (TEM, Joel 200 kV) using ultramicrotomed thin sections (~70 nm thick). Prior to the TEM inspection, butadiene phase was selectively stained by osmium tetroxide (OsO_4).

Results and Discussion

Effect of Chain Architecture on Block Copolymer Morphology

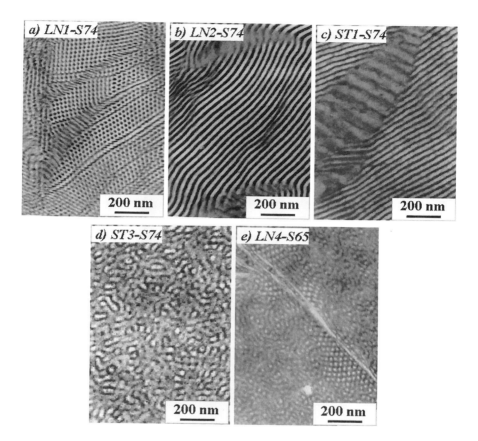

Figure 2.   Representative TEM micrographs of the block copolymers studied; sample notations are given; note that the samples have nearly identical net chemical composition ($\Phi_{PS}\sim$ 0.70).

The equilibrium morphology of the linear block copolymers presented in table 1 have been the object of our previous studies [12,13]. Hence, only a brief account of them will be given, and the star block copolymers are discussed in more detail. Fig 2 illustrates the equilibrium morphology of the block copolymers prepared by solution casting. The micrographs of different samples in Fig 2 (a-e) are arranged in such a way that the variation of morphologies as a result of modified architecture follow in the order: PS matrix (LN1) → Lamellae (LN2, ST1) → bicontinuous-like morphology (ST3) → PS domains (LN4).

It is known from literature that an SB diblock copolymer with about 70 vol.-% PS forms hexagonal polybutadiene (PB) cylinders dispersed in PS matrix [9], and the same will be observed in a symmetric SBS triblock copolymer as shown by TEM image of LN1-S74 in Fig 2a. The change in molecular architecture may completely change the classical picture of morphology formation in block copolymers by overcoming the precondition of altering composition to adjust the morphology.

As demonstrated by our studies, one may notice a wide range of microphase separated structures produced by the variation of block copolymer architectures at constant styrene/butadiene composition (Fig 2). In spite of the net chemical composition equivalent to that of LN1-S74 (Fig 1a) the asymmetric linear block copolymer having a tapered composition profile (LN2-S74 in Fig 2b) and an asymmetric star block copolymer having neat transition (ST1-S74 in Fig 2c) show lamellae. Obviously, the block copolymers having asymmetric architecture exhibit the significant deviation in the composition in which a particular morphology will be formed. The star block copolymer (ST3-S74 in Fig 2d), with 74 vol-% PS and lacking a pure PB as the rubbery phase shows a larger deviation from the classical block copolymer phase diagram. In this star block copolymer, the deviation is substantiated by the presence of a PS-co-PB random copolymer as rubbery block, which contains a considerable part of total polystyrene and hence increases the effective rubber volume fraction. The latter would favour the formation of morphology corresponding to the structure of a diblock copolymer having lower PS content. As a result, a bicontinuous-like morphology is formed. A closer inspection of Fig 2d reveals that the PS struts (grey domains) seem to form an interpenetrating network dispersed in the PS-co-PB matrix. This morphology looks very similar to the 'gyroid' phase observed in block copolymer systems. Finally, at a slightly lower total PS content (65 vol.-% polystyrene) than the rest of the block copolymers, the linear copolymer LN4-S65, which has about 32% by volume of PS as outer pure glassy blocks, forms the glassy domains dispersed in the rubbery matrix [12]. This morphology resembles the structure of classical SBS thermoplastic elastomers, which contain about 30% polystyrene, the composition nearly reverse of LN4-S65.

Table 2. Differential scanning calorimetry (DSC) data of the block copolymers showing the glass transition temperature (T_g) of the soft and the hard phases, heating rate 10°C/min.

Polymer:	T_{g-PB} (soft phase)/°C	T_{g-PS} (hard phase)/°C
LN1-S74	-98	101
LN2-S74	-53	101
LN4-S65	-34	70
ST1-S74	-81	103
ST3-S74	-60	104

It is interesting to note that the change in molecular architecture and hence the morphology formed in the solid state is associated with a shift in the glass transition temperature of the polybutadiene phase (T_{g-PB}), see table 2. With the exception of LN4, where there is a decrease in glass transition temperature of PS phase (T_{g-PS}), the value of T_{g-PS} in other samples remains nearly constant. The rise in T_{g-PB} is a sign of intermixing of polystyrene chains with the polybutadiene phase. The latter depresses the mobility of the PB phase by introducing segments having bulky pendant groups [12]. The elevation of T_{g-PB} in ST3 and LN4 is obvious because of the presence of PS-co-PB random copolymer as rubbery block. The substantial increase of T_{g-PB} even in block copolymers having pure butadiene rubbery blocks (e.g., ST1-S74) suggests that a part of shorter PS chains might have been incorporated into the former. This possible intervention of PS chains, which is augmented by asymmetric conformation, is in line with the recent postulation by Matsen

[14], which allows the increase of soft phase volume fraction in the block copolymers having modified architectures. However, in fact, besides the volume fraction of the phases, the thermodynamics at the interface is the key factor that plays a decisive role in determining the interfacial curvature, i.e., via the interaction parameter of the two species in contact [9]. Thus, a change in chemical nature of the blocks at the interfacial region (neat versus tapered or random) should affect the thermodynamic equilibrium and the resulting morphology.

Effect of Processing Methods on Morphology

The equilibrium morphologies of the block copolymers outlined above are stable ones. These were observed independent of preparation conditions. The processing history may significantly alter the orientation and order of the microdomains [15]. Here, we discuss the influence of sample processing condition taking the lamellar block copolymer LN2-S74 as an example.

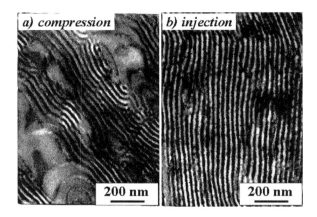

Figure 3. Representative TEM micrographs showing microphase separated morphology of the lamellar block copolymer LN2-S74 in compression mould (left) and injection mould in the middle of the moulded bar (right: injection direction vertical); the equilibrium structure is shown in Fig 2b.

The structures of the block copolymer in compression and injection moulds are qualitatively similar to that found in the solution cast film. The compression-moulded sample shows the well developed lamellar morphology with different orientation of the grains. In Fig 3a, one can notice both edge-on and flat-on views of the lamellar microdomains. The lamellae in the injection mould (Fig 3b) are highly oriented in the flow direction. It should be kept in mind that the morphology of an injection mould can change from one end of the bar to the other and even across its width. In the copolymer studied, especially the continuity of the lamellae was found to be highly affected by the shear force. The orientation of the microstructures along the flow direction is known to cause a pronounced anisotropy in mechanical deformation behaviour [16].

The results discussed so far provide an excellent experimental evidence of possibility of modifying the block copolymer phase diagram via a change in molecular architecture. Thus, the architectural modification of block copolymers may open an interesting way of controlling their morphology and mechanical properties.

Morphology Formation in Block Copolymer/PS Blends

As introduced earlier, the styrene/butadiene block copolymers are often used in combination with polystyrene homopolymer (hPS). It is well known that the length of the homopolymer chains (here hPS chains) relative to that of the corresponding block (here PS block) of the block copolymer plays a vital role in the phase behaviour of the block copolymer/homopolymer blends [9,17].

The low molar mass homopolymer, which is easily assimilated by the corresponding block domains can be used to change the dimension of a given morphology and even to produce new morphologies corresponding to another net composition. Hence, the method of blending with homopolymers can be helpful to locate the position of a 'new' morphology in the phase diagram.

Fig 4 shows the morphology of the star block copolymer ST3-S74 and three different blends with hPS having different molecular weights. In the blends, the hPS weight content is 20% each. The TEM image of ST3 is also included in order to compare the morphology with that of the blends. Three kinds of morphologies can be observed: microphase separated 'bicontinuous' structure (Fig 4a), microphase separated lamellar structures (Fig 4b,c) and macrophase separated morphology containing hPS particles embedded in ST3-S74 matrix.

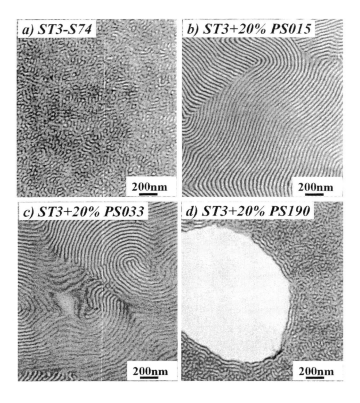

Figure 4.   Representative TEM images of pure star block copolymer ST3-S74 (a), and blends containing 20 wt.-% of homopolystyrene having different molecular weights (b-d).

The polystyrene samples differ in their molecular weights. PS015, PS033 and PS190 have the weight average molecular weight (M_w) of 15 000 g/mol, 33 000 g/mol and 190 000 g/mol, respectively. On the other hand, the molecular weight of the longest polystyrene block in the star block copolymer is in the range of 50 000-70 000 g/mole. Obviously, PS015 and PS033 have shorter chains that can be easily accommodated by the PS blocks of the star molecules. Hence, an addition of 20 wt.-% of PS015 and PS033 leads to the formation of morphology corresponding to higher polystyrene content (lamellar morphology in Fig 4b,c). Since the resulting morphology is a lamellar one (and the type of morphology to be expected for a block copolymer having lower PS content than for the lamellae is 'gyroid' phase), the microphase separated structures observed in the pure ST3-S74 can be considered as being, at least, very close to the 'gyroid' morphology.

Increasing the molecular weight of the added polystyrene (hPS) decreases the entropy of these chains and the hPS chains cannot diffuse into the PS block of the block copolymer. These hPS chains tend to segregate towards the middle of the domains and finally, if the molecular weight is too high, these are even expelled out of the PS domains leading to the formation of macrophase separated hPS particles dispersed in the matrix block copolymer. The molecular weight of the PS190 is obviously much higher than the corresponding block in the block copolymer. Hence the added polystyrene predominantly forms the hPS particles leaving the morphology of the surrounding matrix unchanged (Fig 4d).

The discussion outlined above in case of solution cast samples may not be valid for the mixtures produced by extrusion or injection moulding. The processing may dramatically influence the morphology of the binary block copolymer/hPS blends. Particularly, in ST3-S74/hPS blends, a typical net-work-like structure was formed [23].

Mechanical and Micromechanical Properties

a. Influence of molecular architecture and morphology

Provided that the rubbery PB block is chemically bound to glassy PS blocks, the mechanical behaviour of the phase separated styrene/butadiene block copolymer systems is generally governed by their morphologies. Due to the formation of a wide range of morphologies in a narrow composition range, a great variety of mechanical behaviour was observed in the investigated block copolymers. Fig 5 exemplifies this diversity in linear block copolymers.

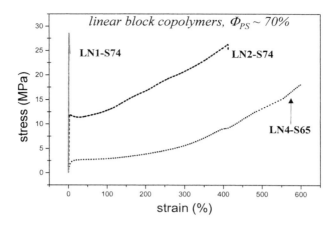

Figure 5. Stress-strain curves of the linear block copolymers using compression-moulded samples according to ISO 527; tensile testing at 23°C and 50 mm/min.

The mechanical behaviour ranges from that of classical SBS thermoplastic elastomer (TPE) to tough thermoplastics and to that of a hard and semi-brittle polymer. This behaviour can be directly correlated to the type of morphologies observed. The TPE character of the sample LN4-S65 is attributed to the morphology comprising dispersed PS domains in the rubbery matrix. Likewise, the highly ductile behaviour of the sample LN2-S74 is a consequence of lamellar morphology. On the other hand, the semi-brittle behaviour of LN1-S74 is correlated with the existence of PS matrix (see Fig 2 for the basic morphologies). The diversity in the mechanical properties are, indeed, the expression of different micromechanical processes, which are discussed in our recent publications [12,18,19]. In the following, we concentrate on the deformation behaviour of a lamellar block copolymer sample (viz., LN2-S74) prepared by different methods and loaded at different temperatures. The micromechanical behaviour of the block copolymers as a function of morphology is summarised in Table 3.

Table 3. Summary of micromechanical mechanisms observed in different block copolymers as a function of morphology (summarised from ref. [12,18,19]).

PS matrix (e.g., LN1-S74)	Alternating lamellae (e.g., LN2-S74)	PS cylinders (e.g., LN4-S65
Crazing with possible void formation in the PS matrix, deformation structures less affected by orientation of PB domains	Yielding of lamellae and fish-bone formation as dominating mechanisms for deformation parallel and normal to the lamellar orientation, respectively	Fragmentation of glassy cylinders, healing up of deformed structures on removing load (a reason for excellent reversibility)

b. Influence of sample processing method

As already indicated earlier, processing might have a dramatic impact on the morphology and mechanical behaviour of block copolymers. Fig 6 plots the stress-strain curves of lamellar block copolymer LN2-S74 processed by three different methods: solution casting, compression moulding and injection moulding. The compression moulded and injected samples met the requirements of standard test (ISO 527) while miniaturised tensile bars (50 mm long, 0.5 mm thick) were prepared from solution cast films. Therefore, the comparison of properties is rather qualitative.

What one can immediately notice in Fig 6 is the significantly high yield stress (ca. 29 MPa) of injection mould. The yield points of solution cast films and compression moulded lie nearly at the same low level (12 MPa). Another noticeable difference is in the tensile strength (maximum stress at break). The tensile strength decreases in the following order: solution cast film $\rightarrow$ compression mould $\rightarrow$ injection mould. It is well known that the tensile strength in styrene/diene block copolymers is a direct function of extent of phase separation [3]. Thus, decreasing tendency of tensile strength in the sequence noted above can be correlated to the decreasing degree of phase separation. The structures in solution cast films are closest to the equilibrium ones, which ensure the largest tensile stress while the injection mould has the smallest chance of forming the well phase-separated morphology owing to the rapid cooling of highly sheared melt. On the other hand, the highest yield stress of the injection mould can be attributed to the orientation of lamellae along the flow direction. Note that the injection-moulded bars were loaded parallel to the injection direction, i.e., parallel to the lamellae orientation direction.

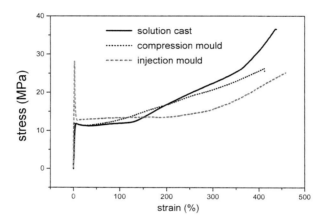

Figure 6.   Stress-strain curves of the lamellar block copolymer (LN2-S74) samples prepared by different methods; injection mould was strained parallel to the injection direction.

The strong influence of processing on the mechanical properties observed indicates that the microdeformation of these samples is governed by different mechanisms. The stress-strain curves (and hence the underlying deformation mechanisms) plotted in Fig 6 can be divided into two categories: a) solution cast and compression mould, and b) injection mould.

In Fig 7, the details of deformation structures observed in solution cast film of the lamellar block copolymer LN2-S74 are presented. The thin section for TEM studies of deformation structures were prepared from tensile bar deformed in the tensile test (sectioning close to the fracture surface). For the purpose of comparison, morphology of this copolymer without deformation is included in Fig 7a.

The TEM micrographs show regions of different structures –lamellae with different forms, thicknesses and long periods. A few of them have been marked as c, d and e (Fig 7b), whose larger magnifications are presented in bottom of Fig 7. The copolymer has originally a long period of 33 nm and a thickness of PS lamellae of 18 nm.

The magnifications in Figs 7a, 7c, 7d and 7e are directly comparable. One can notice that, after deformation to different degrees, the lamellar thicknesses and spacings have been reduced by more than 50% in regions c and d, an indication of a large plastic deformation which arises obviously from the effect of so-called *thin layer yielding* [18].

Due to similarity between the thickness of PS lamellae and that of craze fibrils in bulk homopolystyrene (both in the range of 10-20 nm), the drawing of the PS lamellae can be envisioned as being analogous to the drawing of the craze fibrils, where the PB lamellae in its liquid-like state act comparable to the microvoids and don't hinder the flow-process of the adjacent PS layers. The plastic yielding of glassy PS lamellae, which occurs if the thickness of these layers is below a critical value, is the idea of above mentioned mechanism of *thin layer yielding* [18].

The plastic deformation of glassy lamellae observed in the asymmetric lamellar block copolymer corresponds to the earlier studies of Kawai et al. on unoriented films who showed using TEM and SAXS that during plastic deformation shearing and kinking processes occurs prior to the disruption of glassy lamellae [20,21]. At higher strains, formation of 'fish-bone structure' ('chevron' morphology [16]) characterised by a four point SAXS pattern was observed. Recently Cohen and co-workers reported that during deformation parallel and perpendicular to the lamellar orientation, destruction of lamellar structure and formation of 'chevron' morphology, respectively occur [16].

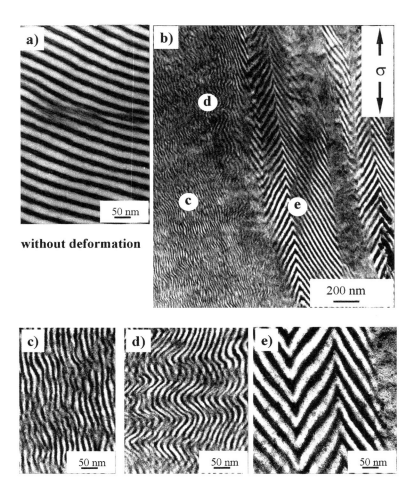

Figure 7.   TEM micrographs showing the morphology of solution cast LN2-S74 without (a) and
after (b-e) deformation; the picture in (a) is for comparison with that in (c,d,e), which
represent the higher magnifications of regions indicated in Fig 7b; strain direction
vertical.

It is interesting to note that undulations with waves are formed in highly drawn lamellae in Fig

7d. These waves appear to evolve by snapping back of the deformed lamellae by release of elastically stored energy after breaking of the tensile specimen. Due to 'polygranular' nature of the solution cast sample, the regions with lamellae perpendicular to the deformation direction lead to the formation of 'chevron'-like morphology. Note that the lamellae in Fig 7e have larger spacing than in the original sample. In consistence with the results of Cohen et al. [16], the lamellar spacing is markedly higher in the 'hinge' of the chevron morphology due to higher dilation of the rubbery PB layers. The increase in the PS lamella thickness further suggests the rotation of lamellar axis during tensile loading.

The strain induced structural changes developed in compression and injection moulded samples provide additional insight into deformation mechanism of the lamellar block copolymer (Fig 8). In the compression mould, there are manifold deformation structures. There are regions, where thickness and periodicity of lamellae changed to different extent after deformation, which represent the different states of plastic deformation of lamellae. In some regions, the lamellae appear to be similar to that of undeformed sample. These might have been formed by rotation of lamellae, which were originally not aligned along the deformation direction, and were not necessarily subjected to drawing. There are even regions, where the PB lamellae are much wider than the PS ones. This can be attributed to the separation of lamellae from each other if these lie perpendicular to the strain direction. On account of the lower Young's modulus and very low glass transition temperature, the polybutadiene lamellae are strongly stretched (dilated) forming isolated regions where PB lamellae (dark) are much thicker than in the undeformed sample (compare Fig 8a with Fig 3a).

By loading injection moulded samples along the flow direction (Fig 8b; compare with Fig 3b), more uniform deformation structures were observed at large deformation. The ordered lamellae structures of the block copolymer has been completely destroyed by plastic deformation. One can estimate the local draw ratio of the PS lamellae of more than 4, suggesting that the local strain of several hundred percent can be achieved by the drawing of glassy phase alone. The yielding of glassy layers as a cause of increased ductility has been discussed in detail in [18]. In the oriented injection moulded sample, the dominant deformation mechanism is, hence, the homogeneous plastic yielding of the layers causing a destruction of an original well-ordered lamellar morphology. On the other hand, in 'polygranular' solution cast films or compression moulds, the principal deformation mechanism at large strains comprise the rotation of the lamellae and lamellar grains towards the strain axis, the shearing in the rubbery layers and the plastic drawing of the glassy lamellae. The shearing in the PB lamellae leading to the higher orientation of polybutadiene chains was confirmed recently by means of FTIR spectroscopy [21].

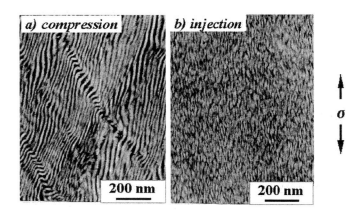

Figure 8. Representative TEM micrographs showing strain induced structural changes in the lamellar block copolymer LN2-S74 in compression mould (left) and injection mould (right); for morphology before deformation see Fig 3; deformation direction is shown by arrow.

c. Influence of temperature

It has been demonstrated that the macroscopic mechanical properties and underlying micromechanical mechanisms in the block copolymer are mainly governed by the nature and alignment of microphase separated structures. So far we have addressed the influence of modified molecular architecture on the morphology formation of block copolymers. We have noted that these modifications are generally associated with a strong T_{g-PB} rise (Table 2), which restrict the lower service temperature of the polymer. Let us examine the deformation behaviour of the lamellar block copolymer LN2-S74 from room temperature down to the temperature close to the glass transition temperature of the soft phase, $T_{g\text{-PB}}$ (details will be available in a future publication [22]).

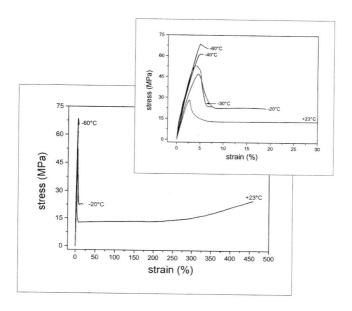

Figure 9.   Stress-strain curves of the injection moulded lamellar block copolymer LN2-S74, loaded at different temperatures as indicated at a rate of 50 mm/min; initial part of the curves are magnified in the inset.

Representative stress-strain curves of LN2-S74 measured at different temperatures are plotted in Fig 9. With decreasing temperature, as expected, the strain at break decreases while the yield stress (a measure of strength) and Young's modulus (a measure of stiffness) increases. At a temperature $\leq$ –40 °C the block copolymer shows a brittle behaviour. Above –40°C, a ductile failure accompanied by a necking of the tensile specimens occurred. In spite of the neck formation and ductile nature of deformation, the strain at break was drastically reduced on lowering the test temperature: e.g., from about 450% (at +23°C) to about 20% (at –30°C). In contrast, the stiffness and strength increase almost linearly with decreasing temperature. The steep decrease in ductility of the lamellar block copolymer at lower temperatures, which is caused by the inability of the neck to extend over the large volume of tensile bar, is indeed an undesirable property.

As indicated previously, a massive increase in the $T_{g\text{-PB}}$, which is an indicator of the occurrence of polystyrene chains in the polybutadiene phase, should have a correlation with a drastically decreased strain at break. The PS chains that reside inside the PB phase seem to stiffen the rubbery layer to such an extent that the PB lamellae can hardly undergo shearing at lower temperatures. Thus, at the vicinity (~-50°C) and below of $T_{g\text{-PB}}$, the sample breaks in a brittle manner.

The ductile nature of deformation at lower temperatures is also manifested in the TEM micrographs of the lamellar block copolymer LN2-S74 deformed at −30°C. The thin sections for the TEM were prepared from a location very near to the fracture surface. The results are presented in Fig 10. The lamellae at −30°C have been plastically drawn homogeneously over a large specimen area. The deformation structures at −30°C and +23°C are qualitatively equivalent to each other. In the necked region, no craze-like zones were observed either.

In spite of the similarity of deformation mechanism at room temperature and at −30°C, the samples showed drastic reduction in the macroscopic strain at break. Due to a localisation of deformation in a limited area of the neck. The sample underwent fracture before the neck could extend macroscopically over the whole tensile specimen. Possible reason for the inability of the neck extension could be found in the fact that the test temperature approaches the glass transition of the soft phase. Therefore, a gliding of the PS lamellae between the soft parts is limited and local dangerous stress concentration can appear. However, it is remarkable that a a homogeneous drawing of the glassy layers (PS) even at the temperature 130°C below its glass transition temperature is possible.

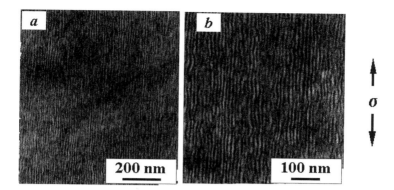

Figure 10. Lower (left) and higher (right) magnifications of TEM micrographs showing strain induced structural changes in injection moulded lamellar block copolymer LN2-S74 at −30°C; deformation direction is shown by arrow.

Summary and Outlook

We have studied the correlation between morphology and micromechanical properties of selected styrene-rich styrene/butadiene block copolymers with respect to their molecular architecture by transmission electron microscopy and uniaxial tensile testing. In particular, the structure-property correlation of an asymmetric linear SBS triblock copolymer was explored by the variation of sample processing methods as well as loading temperature. The micromechanical behaviour was found to be dependent on the alignment of microphases studied. Even at a temperature slightly above the soft phase T_g and far below the hard phase T_g, the glassy polystyrene lamellae showed a ductile behaviour attributable to the nano-scale morphology of the block copolymer systems. In future, systematic studies should concentrate to precisely elucidate how the block copolymer architecture affects the phase behaviour and micromechanical properties of their blends with homopolymers.

196

Acknowledgements

The research was supported by the Kultusministerium des Landes Sachsen-Anhalt (Project: 'Neue Funktionswerkstoffe auf der Grundlage schwach entmischter Blockcopolymere'). We thank Prof. W. Grellmann (University of Halle) for enabling tensile testing. RA is acknowledges research scholarship from Max-Buchner-Forschungsstiftung (MBFSt 6052) and helps from Dr. R. Lach and Dr. R. Weidisch.

[1]   T. P. Lodge, *Macromol. Chem. Phys.* **2003**, *204*, 265.
[2]   F. S. Bates, G. H. Fredrickson, "Block copolymer thermodynamics: Theory and experiment", in: *Thermoplastic Elastomers*, 2nd Edition, G. Holden, N. R. Legge, R. P. Quirk and H.E. Schroeder Eds., Hanser Publishers, Munich 1998, p. 336.
[3]   G. Holden, *"Understanding Thermoplastic Elastomers"* Carl Hanser Verlag, Munich 2000.
[4]   K. Knoll, N. Nießner, *Macromol. Symp.* **1998**, *132*, 231.
[5]   K. Knoll, N. Nießner, "Styroflex – A new transparent styrene-butadiene copolymer with high flexibility", in: *ACS Symp. Series 696, Applications of Anionic Polymerization Research*, R. P. Quirk Ed., p. 112, American Chemical Society 1998.
[6]   K. Geiger, K. Knoll, M. Langela, *Rheol. Acta* **2002**, *41*, 345.
[7]   V. Abetz, T. Goldaker, *Macromol. Rap. Commun.* **2000**, *21*, 16.
[8]   F. S. Bates, G .H. Fredrickson: AIP *Physics today* **1999**, *2*, 32.
[9]   I. W. Hamley, *"The Physics of Block copolymers"*, Oxford Science Publications, Oxford 1998.
[10] C. Lee, S. P. Gido, Y. Poulos, N. Hadjichristidis, N. B. Tan, S. F. Trevino and J. W. Mays, *J. Chem. Phys.* **1997**, *107*, 6460.
[11] S. T. Milner, *Macromolecules* **1994**, *27*, 2333.
[12] R. Adhikari, G. H. Michler, T. A. Huy, E. Ivankova, R. Godehardt, W. Lebek, K. Knoll, *Macromol. Chem.Phys.* **2003**, *204*, 488.
[13] R. Adhikari, R. Godehardt, W. Lebek, R. Weidisch, G. H. Michler, K. Knoll, (**2001**), *J. Macromol. Sci.: Polym. Phys.* **2001**, *40*, 833.
[14] M. W. Matsen, *J. Chem. Phys.* **2000**, *113*, 5539.
[15] R. Adhikari, Ph. D. Thesis, Martin Luther University, Halle-Wittenberg, Germany 2001, http://sundoc.bibliothek.uni-halle.de/dis-online/o1/02H046/of_index.htm
[16] Y. Cohen, R. J. Albalak, G. J. Dair, M. S. Capel, E. L. Thomas, *Macromolecules* **2000**, *33*, 6502.
[17] H. Hasegawa, T. Hashimoto, in: *"Comprehensive Polymer Science, Suppl. 2"*, S. L. Aggarwal, S. Russo Eds., p. 497, Pergamon, London 1996.
[18] G. H. Michler, R. Adhikari, W. Lebek, S. Goerlitz, R. Weidisch, K. Knoll, *J. Appl. Polym. Sci.* **2002**, *85*, 683.
[19] M. Fujimora, T. Hashimoto, H. Kawai, *Rubber Chem. Technol* **1998**, *51*, 215.
[20] T. Hashimoto, M. Fujimora, K. Saito, H. Kawai, J. Diamant, M. Shen, in: *"ACS Advances in Chemistry SeriesMultiphase Polymers"* S. L. Cooper, G. M. Estes Eds., **1979**, *176*, 257.
[21] T. A. Huy, R. Adhikari, G. H. Michler, *Polymer* **2003**, *44*, 1247.
[22] R. Adhikari, G. H. Michler, K. Knoll, *J. Appl. Polym. Sci.* **2003**, in preparation.
[23] Unpublished results

PE film as substrate. The polymerization conditions such as temperature, time, oxidation potential and the concentration of the oxidative agent were optimized in extensive preliminary experiments. The characteristics of the composite film were measured by wide angle X-ray diffraction (WAXD), small angle X-ray scattering (SAXS), scanning electron microphotography (SEM), different scanning calorimetry (DSC), and Fourier transform infrared spectroscopy (FTIR).

Experimental

Sample Preparation: The polyethylene samples used in this experiment were UHMWPE (Hercules 1900/90189) with a viscosity-average molecular weight ($\overline{M}v$) of 6×10^6. The solvent was decalin. The concentration of UHMWPE chosen was 0.4 g/100ml decalin, which is the optimum concentration ensuring the greatest draw ratio of UHMWPE films with $\overline{M}v = 6 \times 10^6$ [11, 12]. Decalin solutions were prepared by heating a well-blended polymer-solvent mixture at 135 ℃ for 40 min under nitrogen. The solution was stabilized with 3 wt% of an antioxidant (di-t-butyl-p-cresol) against UHMWPE. The hot homogenized solution was quenched to room temperature by pouring it into an aluminum tray, thus generating a gel. The decalin was evaporated from the gels under ambient conditions. The resulting dry gel film was vacuum-dried for 24 h to remove any residual trace of decalin, and then elongated up to the desired fold at 135 °C under nitrogen. The thickness of the ultradrawn PE film was in the range of 3~20 μm. A more detailed method has been described elsewhere.[16,17] The ultra-drawn films were washed in ethanol and dried before being used as substrates.

The electric conductivity of PPy polymerized under electric field is well known to increase with decreasing polymerization temperature. This well-known method, cannot however be applied to the polymerization on polyethylene as substrate, since polyethylene is a typical insulator. Accordingly, the ultra-drawn polyethylene film was dipped in purified pyrrole monomers, and transferred into an aqueous $FeCl_3$ solution. The latter was used as an oxidant, in which Cl^- acts as a dopant anion to PPy. [18] After the polymerization, the film was washed with distilled water to remove the excess oxidant, and then with ethanol and acetone to remove the unreacted monomers and oligomers. The resultant film was then dried under vacuum at room temperature for 24 h.

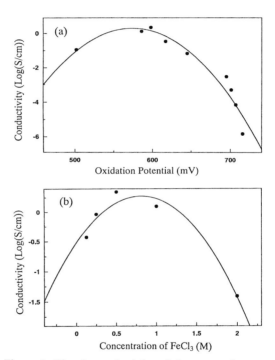

Figure 1. Electric conductivity of the composites as a function of (a) oxidation potential of the solution, in which PPy was polymerized at 20 ℃ for 3 hr in the 0.5 M $FeCl_3$ solution, (b) $FeCl_3$ concentration in the solution with oxidation potential of 600 mV.

To obtain the maximum value of electric conductivity of the PE-PPy composite, the optimum polymerization conditions of PPy on the substrate of PE with $\lambda = 100$ must be determined by controlling four parameters; polymerization temperature, polymerization time, concentration of FeCl$_3$ solution, and the oxidation potential of the solution, apart from the well-known best polymerization conditions of PPy under electric field. According to the Nernst's equation[19], indicating that oxidant potential decreases as Cl$^-$ increases in a FeCl$_3$ solution, a solution with the desired initial oxidation potential was prepared by adjusting the

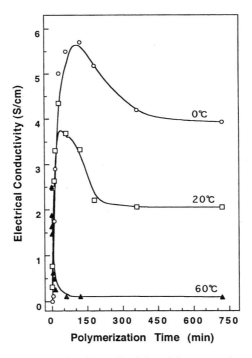

Figure 2. Electric conductivity of the composites vs. polymerization time of PPy at 0, 20, and 60 °C.

amount of Cl⁻ in the solution, namely by adjusting the proportion of HCl and H_2O, before the polymerization and the oxidation potential was measured with a digital pH meter, model HM-20E, provided by TOA Electronics Ltd., Japan. First, polymerization temperature and polymerization time, and the concentration of the solution were fixed to be 20 ℃, 3 h, and 0.5 M. The oxidation potential of the solution was changed from 500 to 720 mV. Figure 1(a) shows the results. The conductivity of the resultant composite becomes the maximum at about 600 mV.

In the next step, the conductivity of the composite was obtained as a function of the concentration of $FeCl_3$ in the solution with an oxidation potential of 600 mV. The concentrations were varied in the range from 0.1 to 2 M and the polymerization temperature and the polymerization time were also fixed at 20 ℃ and 3 h, respectively. Figure 1(b) shows the change in electric conductivity against $FeCl_3$ concentration. The electric conductivity of the composite was found to be maximum at 0.5 M concentration. By using the solution with 0.5 M concentration and with oxidation potential of 600 mV, the composites were prepared by changing the other two parameters, polymerization temperature and polymerization time.

Figure 2 shows the electric conductivity of the composites against polymerization time at 0, 20, and 60 °C. This experiment was carried out to check the relationship between the amounts of polymerized PPy on polyethylene film and the electric conductivity. In the initial stage, the electric conductivity increases drastically and reaches a maximum value, indicating a rapid formation of conductive paths. The further increase in polymerization time, beyond each maximum value of conductivity for the three composites, causes a drastic decrease in conductivity and it tends to level off. This tendency is most considerable at 60

°C. The decrease of conductivity of PPy is thought to be due to change in the oxidation potential of the solution with increasing polymerization time. Interestingly, the maximum values and the time elapsed prior to arriving at the maximum are sensitive to the polymerization temperature. Namely, the magnitude of electric conductivity is highest at 150 min, when the polymerization is done at 0 °C. This phenomenon shall be discussed later in relation to the temperature dependence of the chemical structure of PPy, as determined by FTIR and ^{13}C NMR spectra.

Characterization of UHMWPE-PPy composites: The electrical conductivity of the drawn films was measured in the stretching direction by using a two-terminal method at room temperature or in the temperature range from -150 to 220 °C.

The DSC measurements were done with an EXTRA-6000, furnished by Seiko Instrument Inc. Sample sizes of 1 mg were used, under nitrogen, and the heating rate was 10 °C/min. An FTIR-8300 from Shimadzu was used to record the infrared spectra of the composite films.

X-ray measurements were carried out with a 12-kW rotating-anode X-ray generator (Rigaku RAD-rA). Monochromatic CuK α radiation (wavelength of 0.154nm) was used. WAXD patterns were obtained with a flat camera and SAXS intensity distribution in the meridional direction was detected with a position sensitive proportional counter (PSPC). Corrections of X-ray scattered intensity were made for air scattering, polarization and absorption.[20]

SEM photographs were taken with a HITACHI-X650, with a working voltage of 20 kV and magnification 1000.

The temperature dependence of the dynamic tensile modulus was measured with a visco-elastic spectrometer (VES-F), obtained from Iwamoto Machine Co., Ltd., at a fixed frequency of 10Hz over the temperature range of -150-300 °C. The Young's modulus and tensile strength were measured at room temperature with a Tensilon / STM-H-500BP.

Results and Discussion

Figure 3 shows the SEM photographs of the surfaces of the PE-PPy composites prepared at the indicated polymerization temperatures and polymerizetion times. The number of PPy particles on the surface increases and the particle size becomes bigger with increasing temperature and time. After 10 sec, the substrate's surface is covered by a large amount of PPy particles when the polymerization was done at 60 °C. Judging from the results in Fig. 2, it may be expected that the electric conductivity of PPy is sensitive to polymerization temperature. This phenomenon is thought to be due to the dependence of the chemical structure of Ppy on polymerization temperature. This shall be discussed later when the FTIR and ^{13}C NMR spectra are considered.

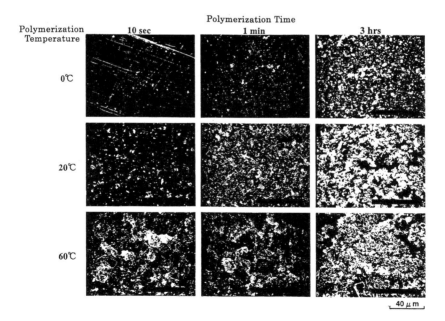

Figure 3. SEM photographs of the surface of PE-PPy composites, in which PPy was polymerized at the indicated polymerization temperature and polymerization time.

Table 1 The density of films and the weight fraction of PPy in the PE/PPy blend films at the different draw ratios

		Draw ratio λ (fold)							
		1		50		100		200	
		Density (g/cm^3)	Weight fraction (%)	Density (g/cm^3)	Weight fraction (%)	Density (g/cm^3)	Weight fraction (%)	Density (g/cm^3)	Weight fraction (%)
PE		0.9768	–	0.9858	—	0.9933	—	0.9998	—
PE-PPy	0℃	0.9833	0.50	1.0821	27.78	1.0522	37.27	1.0497	57.25
	20℃	0.9846	0.69	1.2180	34.51	1.1030	63.12	1.0993	58.04
	60℃	0.9931	3.59	1.2138	52.46	1.1754	65.81	1.1199	64.32

Here we must emphasize that the PPy particles that over-lapped on the surface could be removed by soft paper, since most of polymerized PPy particles were on the surface of PE, they had not penetrated into the interior of PE film, as shown in Fig. 3, at elevated polymerization temperature. This is important, to estimate the amount of PPy that penetrated into cracks between oriented PE fibrils, in measuring electric conductivity. Therefore the specimens with no overlapped particles on the surface were used as test specimens in the following experiments. Table I lists the weight fractions of PPy in PE-PPy composites prepared by the polymerization of PPy at 0, 20, and 60 °C after removing the overlapped particles. The polymerization time was set to be 2 h, assuring the highest conductivity of PPy polymerized at 0 °C. To check the draw ratio dependence on the conductivity of the composites, PE films with different draw ratio were used as substrates.

206

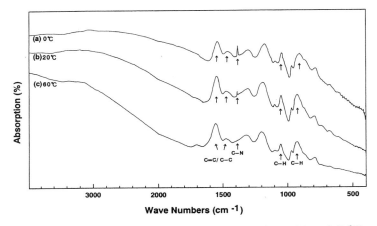

Figure 4. FTIR spectra measured for PPy polymerized at 0, 20, and 60 °C.

The weight fraction was calculated by the weight of the film before and after the polymerization. The PPy fraction of the composite becomes higher as the draw ratio of PE as substrate becomes higher. This is because of the decrease of film thickness with increasing draw ratio. The cross-section area was calculated from the density, measured using a pycnometer, with toluene and carbon tetrachloride as solvents. Direct measurement of the film thickness using a micrometer was impossible, since the thickness was too thin to be estimated accurately. Accurate values of the cross-section area are very important to estimate Young's modulus and tensile modulus of the composite films.

Returning to Fig. 2, it is seen that the electric conductivity of PPy is lower with increasing polymerization temperature, although the amount of polymerized PPy increases. To explain this phenomenon, FTIR spectra were measured for PPy polymerized at 0, 20 and 60 °C. Figure 4 shows the results. The N− H or C− H feature absorption is invisible from 1600 to 4000 cm⁻¹, since they are masked by the tail of the 1 eV peak generally formed due to the doping.[21] In the region from 1600 to 400 cm⁻¹, the vibrations of pyrrole ring appear, i.e. at 1540 and 1470cm⁻¹, associated with the C− C and C=C stretching, and at 1384 cm⁻¹, associated with C− N stretching. According to the quantum chemical calculation by Tian et

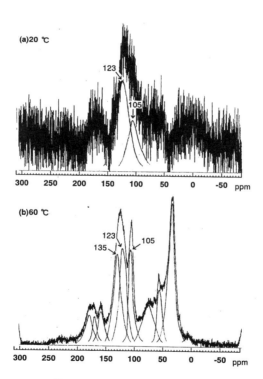

Figure 5. ¹³C NMR spectra of PPy polymerized at 20 and 60 °C.

al.[22, 23], the dopant molecules will lead to the abovementioned bands being selectively and strongly enhanced. In our IR spectra, the PPy polymerized at lower temperature is considered to be of higher doping degree. Following Tian and Zerbi[21-24] the ratio of the absorption intensity of 1550 and 1470 cm^{-1} (I_{1550}/I_{1470}) was reported to be inversely proportional to the extent of delocalization, namely, the conjugated length along the polymer chains. In Figure 4, the I_{1550}/I_{1470} is 2.6, 3.0, and 4.0 for the PPy prepared at 0, 20 and 60 °C, respectively, indicating that PPy synthesized at lower temperature has a lower I_{1550}/I_{1470} value. This means that PPy polymerized at the lower temperature has longer extended conjugation and less defect sites in the chain than the other two.

The solid ^{13}C NMR spectra for the PPy polymerized at 20 and 60 °C are shown in Figure 5. The conductive samples were dispersed by grinding together with salt powder before setting into the sample rotor. The spectrum of PPy polymerized at 0 °C could not be obtained, while that at 20 °C contained much noise. These phenomena are attributed to electric discharge in the magnetic field, as frequently occurs with conducting materials. Both spectra 5(a) or 5(b) show a very broad asymmetric peak shifted downfield, compared with neutral PPy.[25] The major peaks at 123 and 105 ppm downfield from tetramethylsilane (TMS) correspond to the α- and β-carbons of the pyrrole monomer, respectively, which appear normally at 117 and 108 ppm relative to TMS.[21, 26] The peak at 135 ppm may indicate the presence of some non α– α linkages, such as α– β linkage or chain end groups. The PPy synthesized at 20 °C shows more linkage of α carbon than PPy synthesized at 60 °C. This is in good agreement with the structural analysis for PPy, indicating that the higher conductivity PPy has more α– α linkages. [21] The defect-forming reaction with high activation energy is predominant in comparison with the desired α– α linkages reaction as the polymerization temperature increases. Since the defect-forming reaction causes cut-off spots in the conjugating chain and induces irregularity in molecular structure, PPy polymerized at higher temperatures leads to low electric conductivity.

Figure 6 indicates the temperature dependence of the electric conductivity of PE ($\lambda = 200$) and PE/PPy ($\lambda = 200$) composite films prepared by using PPy polymerized at 0, 20 and 60 °C. The polymerization time was fixed at 2 h (based on the results in Fig. 2). Conductivity measurements of the composites were done in temperature range from −150 to 150 °C in air. The conductivity of the composite is much higher than that of the PE film. As shown in Fig. 2, the conductivity of the composite film prepared at polymerization temperature 0 °C is higher than that of the samples prepared at the other polymerization temperatures, 20 and 60 °C.

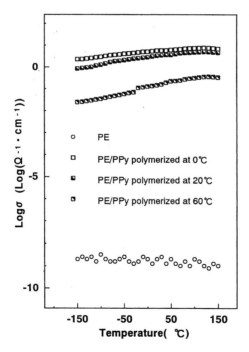

Figure 6. Temperature dependence of electric conductivity of PE and PE/PPy composite films prepared by using PPy polymerized at 0, 20, and 60 °C.

The increase in conductivity with temperature is very small. This means that the electrical property of the composite film is stable in air in the temperature range from −150 to 150 °C. Incidentally, the draw ratio of the composite films is almost independent of the electric conductivity in the given temperature range. Then we shall discuss the mechanical properties of the composite films as a function of draw ratio.

Figures 7 show the Young's modulus and the tensile strength for PE films and PE/PPy composite films.

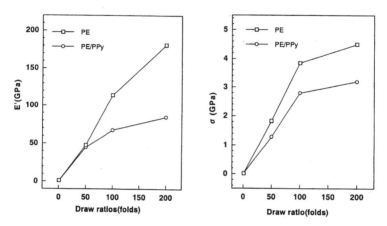

Figure 7.   Young's modulus and tensile strength of PE films and PE/PPy composite films as a function of draw ratio.

The polymerization of PPy was carried out in the $FeCl_3$ solution at 0 °C for 2 hrs, which corresponds to the optimum condition as discussed before.   It is clear that the higher the draw ratio of PE as substrate, the higher the Young's modulus and the tensile strength of the

PE/PPy composite film. For the PE film with 200 folds, Young's modulus and tensile strength reach 180 and 4.5 GPa, respectively, while Young's modulus and tensile strength of the corresponding composite film reach 80 GPa and 3.2 GPa. The drastic decrease is due to the fact that there is no contribution of PPy to the mechanical properties, Young's modulus and tensile strength and this is attributed to the increase in the cross section area of the composite films, as shown in Table 1. We must emphasize that the Young's modulus of the composites is practically the same as that of drawn PE homo-polymer films within the experimental error. Because, Young's modulus of PPy is negligible in comparison with the Young's modulus of PE and the introduction of PPy only increases the cross section area of the composite.

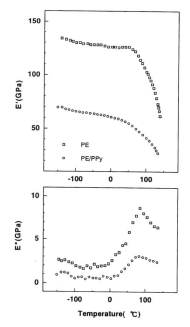

Figure 8. Temperature dependence of the storage and loss moduli of the PE film and PE/PPy composite film at λ =100.

Figure 8 shows the temperature dependence of the storage and the loss moduli of the PE film and PE/PPy composite film with $\lambda = 100$. The polymerization temperature and polymerization time were selected to be 0 °C and 2 hrs, respectively. The decrease in the storage modulus for the composite film compared with that of the PE film is also due to the increase in the cross section area by the introduction of PPy as discussed above. The loss modulus showed the α-dispersion peak associated with crystal dispersion[27-28] of PE appears around 90 °C for both the specimens. The β–mechanism, which appears around – 30 °C – 50 °C generally for melt PE films and is associated with the amorphous dispersion, is not observed for the composite because of high crystallinity of the ultradrawn PE matrix. Accordingly, the temperature dependence of the storage and loss moduli indicates that compositing of PPy caused no damage to the crystalline within PE substrate.

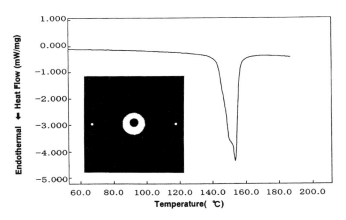

Figure 9. DSC and X-ray diffraction pattern for PE/PPy composite (λ =100).

To confirm this conclusion, DSC and X-ray diffraction measurements were done for the PE film and PE/PPy composite film with λ= 100. Figure 9 shows the results. DSC curve shows an endotherm peak at 152 °C for the PE/PPy composite film. The curve with same profile was also observed for the ultradrawn PE film.[16] The WAXD pattern shows strong equatorial reflecting spots from the (110) plane, as observed for the ultradrawn film, indicating a high degree of orientation of the c-axes with respect to the stretching direction. Judging from the WAXD pattern, the introduction of PPy particles causes no effect on the high orientation of the PE crystallites, since PPy is of extremely poor crystalline, as reported for all forms of PPy so far.

WAXD pattern and SAXS patterns were taken from edge view for the undrawn composite film, although they are not shown in this paper. The reflection from (110) and (200) planes showed the preferential orientation of the c-axes perpendicular to the film surface from the gelation process. The SAXS pattern showed the scattering maxima in the meridional direction, indicating that the crystal lamellae were oriented with their flat faces parallel to the film surface. The both patterns are the same as the PE undrawn films prepared by gelation/crystallization from solutions.

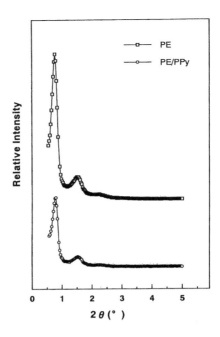

Figure 10. SAXS intensity distribution as a function of twice Bragg angle (2θ) in the direction parallel to the film surface for the undrawn PE and PE/PPy .

In order to obtain the precise information, SAXS intensity in the meridional direction parallel to the film surface (end view) was observed as a function of twice the Bragg angle (2θ) by PSPC system for the undrawn PE and PE/PPy composite films. The result was shown in Figure 10. Typically, undrawn UHMWPE dry gel film shows a SAXS profile reflecting alternating lamellae with an identity period. [29] The three meridional scattering maxima indicate that the film is composed of highly oriented crystal lamellae with their faces parallel to the film surface. For the PE/PPy composite film, the strength of the scattered maxima become indistinct, the peaks can be observed at the same twice the Bragg angle. Thus it turns out that the orientation of the crystal lamellae with their faces parallel to the film surface causes no significant by the introduction of PPy particles. The change in the intensity is thought to be due to the absorption of X-ray by PPy

Conclusion

A conductive PE/PPy composite film was synthesized by chemical polymerization of a pyrroles monomer on an ultradrawn PE substrate. The optimum reaction conditions, were the following: polymerization temperature 0 °C, polymerization time < 2 h, concentration of $FeCl_3$ in the aqueous solution 0.5 M, and oxidation potential was about 600 mV. The Young's modulus and tensile strength of the resultant composite film prepared from PPy synthesized under these optimum conditions reached 80 GPa and 3.2 GPa, respectively, and the electrical conductivity reached 5.5 S/cm at room temperature. No damage to the oriented PE substrate by the introduction of Ppy was confirmed. This lead to the successful production of conductive polymers with high modulus and high strength.

[1] J. P. Riggs, A. Harris, and J. B. Stamatoff, Frontiers of Polymer Research, P. N. Prasad and J. K. Nigam, Ed., Plenum Press, New York, **1991**. p.27-44.

[2] G. B. Street, Handbook of Conducting Polymers, T. A. Skotheim, Ed., New York, M. Dekker, **1986**, p.265-291.

[3] K. Kanazawa, et al., *Synth. Met.* **1979**, *1*, 329.

[4] M. Salmon, K. Kawakawa, A. F. Diaz, and M. Krounbi, *J. Polym. Sci. Polym. Lett. Ed.,* **1982**, *20*, 187.

[5] A. Diaz, and B. Hall, *IBM J. Res. Dev.* **1983**, *27(4)* 342.

[6] S. E. Lindsey and G. B. Street, *Synth. Met.,* **1984**, *10*, 67.

[7] O. Niwa and T. Tamamura, *J. Chem. Soc., Chem. Commu* , **1984**, *13,* 817.

[8] M. Omastova, I. Chodak, J. Pointech, *Synth. Met.* **1999**, *102,* 1251.

[9]   P. Smith, P. J. Lemstra, B. Kalb, A. J., Pennings, *Polym Bull*, **1979**, *1* 733.

[10]   P. Smith, P. J. Lemstra, J. P. L. Pijpers, A. M. Kiel, *Colloid & Polym. Sci.*, **1981,** *259,* 1070.

[11]   M. Matsuo, K. Inoue, N. Abumiya, *Sen-I Gakkaishi*, **1984,** *40*, T275.

[12]   M. Matsuo, C. Sawatari, *Macromolecules*, **1986,** *19*, 2028.

[13]   C. Xu, Y. Agari, M. Matsuo, *Polym J.*, **1998,** *30(5)* 372.

[14]   T. Kanamoto, A. Tsuruta, K. Tanaka, M. Katada, R. S. Portor, *Polymer J.*, **1983,** *15,* 327.

[15]   T. Kanamoto, T. Ohama, K. Tanaka, M. Takeda, R. S. Portor, *Polymer* **1987,** *28*, 1517.

[16]   T. Ogita, R. Yamamoto, N. Suzuki, F. Ozaki, M. Matsuo, *Polymer* **1991,** *31*, 882.

[17]   M. Matsuo, C. Sawatari, M. Iida, M. Yoneda, *Polymer J.*, **1985,** *17,* 1197.

[18]   A. Techagumpuch, Frontiers of Polymer Research, P. N. Prasad and J. K. Nigam, Ed., Plenum Press, New York, **1991,** p351-357.

[19]   C. X. Fu, W. X. Shen and T. Y. Yao, *Physical Chemistry*, High Education Press: **1993**, p 505.

[20]   G. D. Zhou, L. Y. Duan, The Structural Chemistry, Beijing Uni. Press, **1995**.

[21]   J. E. Mark, Ed., Polymer Data Handbook, Oxford University Press, New York, **1999**.

[22]   B. Tian, G. Zerbi, *J. Chem. Phys.*, **1990,** *92(6),* 3886.

[23]   B. Tian, G. Zerbi, *J. Chem. Phys.*, **1990,** *92(6),* 3892.

[24]   T. A. Skotheim, R. L. Elsenbaumer, J. R. Reynolds, Ed., Handbook of Conducting Polymers (2nd Edition), Marcel Dekker, New York, **1998**, p.409-421.

[25]   G. B. Street, T. C. Clark, M. Krounbi, K. Kanazawa, V. Lee, P. Pfluger, J. C. Scott, G. Weiser, *Mol. Cryst. Liq. Cryst.*, **1982,** *83*, 253.

[26]   T. C. Clark, J. C. Scott, G. B. Street, *IBM J. Res. Dev.*, **1983,** *27,* 313.

[27]   C. Sawatari, and M. Matsuo, *Coll. & Polym. Sci.* **1985,** *263,* 783.

[28]   M. Matsuo, C. Sawatari, and T. Ohhata, *Macromolecules* **1988,** *21*, 1317.

[29]   M. Matsuo, M. Tsuji, R. St. J. Manley, *Macromolecules* **1983,** *16,* 1.

Macromol. Symp. **2004**, *214*, 217-230

Use of the Surfmer 11-(Methacryloyloxy) undecanylsulfate MET as a Comonomer in Polystyrene and Poly(methyl methacrylate)

P. C. Hartmann,[1] *A. Pienaar,*[*1] *H. Pasch,*[*2] *R. D. Sanderson*[1]

[1] UNESCO Associated Centre for Macromolecules, Department of Chemistry and Polymer Science, University of Stellenbosch, Private Bag X1, Matieland, 7602, South Africa
[2] Deutsches Kunststoff-Institut, Schloßgartenstr. 6, D-64289 Darmstadt, Germany

Summary: The polymerizable surfactant sodium 11-(methacryloyloxy) undecanylsulfate (MET) has been synthesized with high purity, and its thermal stability and phase transitions have been studied by thermo gravimetric analysis (TGA) and differential scanning calorimetry (DSC), respectively. MET has been copolymerized in solution with methylmethacrylate (MMA) or styrene (S), initiated by azo-bis-isobutyronitrile (AIBN). The copolymers thus obtained have been studied by Gel Permeation Chromatography (GPC), Transmission Electron Microscopy (TEM), and DSC. Due to the incompatibility between the polar head of the MET units and the non polar S or MMA units, MET units organize in the amorphous polymer matrix and arrange in lamellar structures.

Keywords: copolymerization; differential scanning calorimetry (DSC); polymerizable surfactant; TEM

Introduction

"Surfmers" (polymerizable surfactants) behave like classical surfactants and self organize in aqueous solution prior to polymerization. Since the first attempts to polymerize polymerizable surfactants in aqueous media by Sherrington[1, 2] (which led to the formation of polysoap rather than polymerised micelles), surfmers have attracted considerable attention, mainly in emulsion polymerization and for templating polymerization[3-9]. In emulsion polymerization, for example, they reduce most of the disadvantages associated with the use of classic surfactants, such as flocculation, high permeability of polymer film and water retention, while advantageously increasing the stability of the latex during storage. Thus, surfmers

DOI: 10.1002/masy.200451015

have been used for the preparation, in microemulsion, of size-controlled polymer particles on different scales (micro to nanometer)[10-19]. Surprisingly, only a few authors have reported on the organization of surfmers in bulk when copolymerized with a non surface-active comonomer. Favresse et al[20, 21] described the formation of super-structures in bulk with polymerized or copolymerized zwitterionic surfmers: poly(carbobetaine)s can form lamellar or interdigitated structures, depending on their chemical structure. When copolymerized with N-vinylpyrrolidone or dimethylacrylamide, carbobetaines based on isobutylene form lamellar phases in bulk[20]. In this last example, the copolymers contain a high ratio of surfmers (about 45 to 70 mol%).

To our knowledge, there are no reports in the literature on copolymers containing low ratios of surfmer units that form supramolecular structures in bulk (eg. lamellar structures). In this work, we prepared copolymers of styrene (S) and copolymers of methylmethacrylate (MMA) containing a low ratio of the surfmer sodium 11-(methacryloyloxy) undecanylsulfate (MET). We chose MET as the surfmer because its synthesis is known, it is readily polymerizable, and copolymerizable with S or MMA.

Pursuing the objective to determine the role played by MET in the organization in bulk when copolymerized with S or MMA, we describe here the synthesis and the properties of PS and PMMA containing 1.5 to 6 mol% of MET units. Properties were determined by differential scanning calorimetry (DSC), thermo gravimetric analysis (TGA) and transmission electron microscopy (TEM). The thermal behaviour of the polymerizable monomer MET was studied by DSC, TGA, and microscopy under cross-polarized light.

Experimental

Azo(bis)isobutyronitrile (AIBN) was recrystallized from methanol. S and MMA were distilled prior to use. MET was synthesized according to a procedure described in the literature[22].

GPC analysis of the copolymers was performed on a Waters 510 HPLC. A PL-ELSD 1000 (light scattering) detector and a PL-gel Mixed A column were used. A sample concentration of 1 mg/ml and eluent flow of 1 ml/min were used. All measurements were performed at 50 °C in dimethylformamide (DMF). Differential scanning calorimetry (DSC) was done using a DSC22C (Seiko II), at a heating rate of 10 °C/min.

Copolymerizations were done by dissolving 1 wt% AIBN in chloroform at 70 °C. 2.0g (0.019 mol) of S or MMA and 5, 10, or 20 wt% MET were dissolved in chloroform and added dropwise over 10 min to the solution of the initiator. The reaction mixture was then refluxed for 4 h at 70 °C. The polymer obtained was precipitated in methanol and dried under reduced pressure.

Monomer MET

A convenient approach to the synthesis of a polymerizable surface-active molecule is to attach the polymerizable group to the hydrophobic tail first, leaving the insertion of the head group to the end of the reaction sequence. Using this method, many practical problems frequently associated with isolation and purification of surface-active intermediates are avoided. Methacrylic acid and 11-bromoundecanol were used as starting material for the preparation of 11-(methacryloyloxy) undecanol. Methacrylic acid was firstly deprotonated with sodium hydroxide to give the corresponding sodium acrylate. 11-(methacryloyloxy) undecanol was obtained from the 11-bromoundecanol and sodium methacrylate using phase-transfer-catalyst conditions. The obtained alcohol was then sulfonated with neat chlorosulfonic acid, and the resulting product quenched with sodium carbonate to obtain the MET. Results of ^{1}H-, and ^{13}C-NMR, and elemental analysis showed that the MET was >95 % pure.

Analysis of MET by TGA showed two distinct stages of degradation. The onset T_{on} and maximum T_{max} temperatures of the first step (219.2 and 250.2 °C), and the second step (322.2 and 411.0 °C) are illustrated in Fig. 1. The maximum in the weight loss observed reached 0.89 g/min at T_{max}.

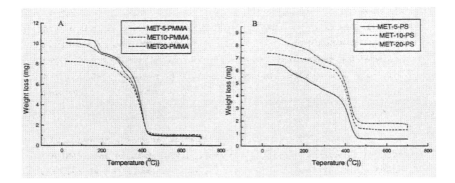

Figure 1.   TGA of the monomer MET

DSC analysis of MET displayed two distinct transitions for the first heating cycle, see Fig. 1, which reveals thermotropic behaviour. Although the precise nature of the mesophase observed has not been determined in this work, we assume that the MET monomer forms a liquid crystalline phase and not Condis-cristals[23, 24] between the two-phase transitions. The crystal to liquid-crystal transition appears at 57.1 °C, followed by the clearing point transition at 85.5 °C. The cooling cycle shows a crystallization transition at T_c=74.7 °C, which cannot be assigned to a specific phase transition due to supercooling and partially to hysteresis attributable to the temperature scan rate. The second heating cycle (see Figure 2) confirms the reproducibility of the two transitions observed in the first heating cycle. There was however a slight shift in the temperatures of the transitions (53.6 and 84.1 °C). In addition, the enthalpies of melting are smaller for the second heating than for the first heating, due to the loss of crystallinity occurring during the first heating cancelling any thermal history (see Table 1). The observation of MET by microscopy under cross-polarized light, with heating stage, confirmed the existence of the mesophase. When the sample is heated from room temperature, a change in crystal texture occurs from 54 to 88 °C. The mesophase is stable from 88 to 107 °C, thereafter it becomes transparent as the temperature increases. The sample is totally isotropic at 152 °C. When cooled, the MET forms a fan-like texture below 110 °C.

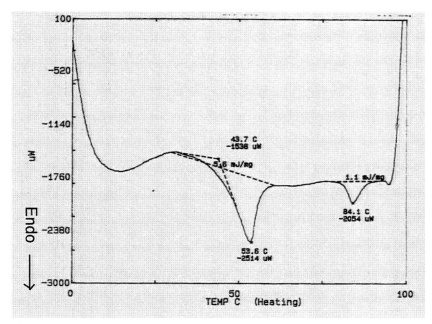

Figure 2. DSC of the monomer MET

Table 1. DSC data of the monomer MET

	Temperature (°C)		Enthalpy (KJ/mol)
First heating	T_m	57.1	15.2
	T_{cl}	85.5	1.2
Cooling	T_c	74.7	-0.5
Second heating	T_m	43.7	2.0
	T_{cl}	84.1	0.4

Copolymers

The copolymerisation of surfmers with non-polar comonomers such as S or MMA is generally performed in emulsion[25-27]. Accordingly, the surfmer MET has been copolymerized in emulsion with various comonomers, including S[17, 22]. Because the objective of our work was to determine the role played by the MET comonomer on the properties of PS and PMMA in bulk, we intentionally synthesized the copolymers by free-radical polymerization in solution and not in emulsion, so as to prevent any self-aggregation of the surfmer prior to polymerization. The solvent used (chloroform) dissolved both comonomers well, and aggregates were not formed by the surfmer. The reactivity ratios of MET with S and with butyl acrylate have been reported by Otsu[28] and Urretabizkaia[29] (see Table 2). From these values (0.53 and 0.30 respectively, for r_S and r_{MET}), one can assume that the MET-S copolymers are random copolymers, with a tendency to alternate. On the other hand, for the copolymers of MMA, we did not deduce the distribution of the comonomers because of the lack of data in the literature about the reactivity ratios of comparable comonomers.

Table 2. Reactivity ratios of MET with styrene and butyl acrylate[22]

	Styrene (S)[28]		Butyl Acrylate (BA)[29]	
Reactivity ratios	r_S	r_{MET}	r_{BA}	r_{MET}
	0.53	0.30	0.32	2.6

After purification of the copolymer by precipitation in methanol, the yields obtained reached only 20 %. The copolymerization products, MET-PS and MET-PMMA, were characterized by FT-IR. All purified samples showed residual amounts of monomer left in the copolymer as determined by the presence of a very small vibration band for the carbon-carbon double bond at about 1650 cm^{-1} in the FT-IR spectrum. The sensitivity of the FR-IR method did however not permit the accurate determination of the ratio of monomer left in the copolymer. Typical infrared spectra that are representative of MET-PS and MET-PMMA are shown in Figs. 3 and 4, respectively. A summary of the assignment of bands representative of both types of copolymers is given in Table 3. The amount of incorporated MET was assessed from the ratio between the integrated areas of the carbonyl band centred around 1720 cm^{-1} and the $-$C-O- band around 1144 cm^{-1} (for copolymers with PMMA) or the aromatic band at 697 cm^{-1} (for copolymers with PS) after correction by "attenuated total reflexion". The copolymers displayed an increase in the ratio of integrated areas as the amount of MET was increased (see Table 4). Because the accuracy of the FT-IR was insufficient, we considered that the ratio of MET in the copolymers is equal to the ratio of the monomer MET fed for the polymerization.

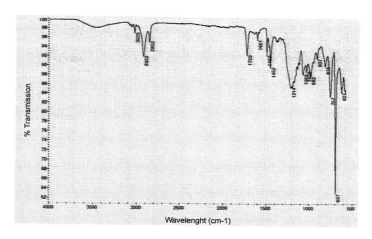

Figure 3. Typical FT-IR spectrum of the MET-PS copolymers

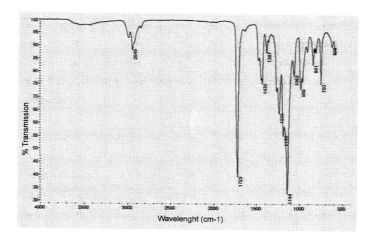

Figure 4. FT-IR spectrum of MET-PMMA

Table 3.   FT-IR band assignments for copolymers MET-PS and MET-PMMA

	Poly(S-co-MET)	Poly(MMA-co-MET)
Assigned group	Wavelength (cm-1)	Wavelength (cm-1)
-CH2-	3026, 2923, 2853	2994, 2949, 2857
-O-C=O	1720	1723
Aromatic ring	1601	-
Aromatic ring	1493	-
-CH2-	1452	1435
-CH3	1387	1387
-C-O-	1213	1239, 1144
-SO2O-	1069	1063
Aromatic ring	757, 697, 627	-
-CH2-	815	841, 750

Table 4.   Ratios of MET monomer units determined from the FT-IR band ratio

copolymer	FT-IR band ratio
MET-1.5-PMMA	0.922
MET-3-PMMA	0.949
MET-6-PMMA	1.041
MET-1.5-PS	0.665
MET-3-PS	0.671
MET-7-PS	1.438

Concerning the solubility of the copolymers: we observed that they remained in solution after the copolymerization reaction, although, after precipitation, the dried copolymers would not re-dissolve in chloroform. This behaviour has been reported before[1, 30, 31], even in the case of polymers, which appear to be initially fully soluble. It seems that in the dry state the ionic groups are tightly bound within a hydrophobic network. Indeed, they may form microphase-segregated domains[1]. The product has trouble dissolving after precipitation and drying, even after a prolonged dissolution time. Copolymers of mixed polarity are known for their solubility problems, as one of the comonomer is very polar and the other is non-polar[30]. MET is soluble in polar solvents, whereas PMMA and PS are soluble in non-polar solvents. Both series of copolymers were insoluble in the following solvents: water, 1,2-dichlorobenzene, 1,2,4-trichlorobenzene, methanol, dichloromethane, chloroform, acetone, 1-methyl-2-pyrrolidone and hexafluoro-2-propanol. Toluene, m-cresol and chlorobenzene only swelled the samples. The Samples were soluble only in tetrahydrofurane (THF) or in DMF after being heated at 80 °C and stirred for 24 h.

The determination of the average molecular weight of the copolymers was performed by GPC in THF. Most of the chromatograms displayed an unusual tri-modal distribution (see Fig. 5 for example). The molecular weights calculated for each copolymer, using a PS calibration curve, are summarized in Table 5.

It was observed that the very high molecular weight fraction has a narrow molecular weight distribution, while the fraction with medium molecular weight has a broad molecular weight distribution. The narrow molecular weight distribution is because the column has an exclusion limit above 10 7g/mol, and macromolecules of bigger size are totally excluded. The lower molecular weight distribution is broader, which is more in accordance with the free radical mechanism of the copolymerization. It can be speculated that the two very high molecular weight distributions arise from the formation of aggregates when the copolymer is in solution. In order to determine whether aggregations occur at these extremely high molecular weights, fractions were taken at the different peak elution times in order to investigate the chemical composition of each fraction. The different fractions were analysed by FT-IR. Fractions of both low and intermediate molecular weights displayed similar spectra to that of the dry copolymer sample. On the other hand, the fraction with the very high molecular weight distribution exhibited an absorption band at 1659 cm-1, indicating the presence of a carbon-carbon double bond. This shows that a residual amount of MET monomer forms micelles or aggregates with the copolymer, giving rise to the presence of the high molecular weights distributions. Moreover, the intermediate molecular weight probably arises from aggregation of copolymers in THF.

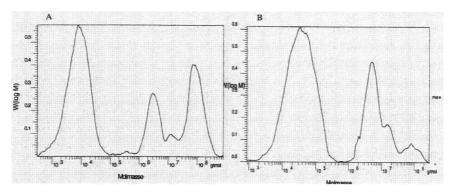

Figure 5. Typical molecular weight distribution for (A) MET-10-PS and (B) MET-10-PMMA

It was observed that the very high molecular weight fraction has a narrow molecular weight distribution, while the fraction with medium molecular weight has a broad molecular weight distribution. The narrow molecular weight distribution is because the column has an exclusion limit above 10^7 g/mol, and macromolecules of bigger size are totally excluded. The lower molecular weight distribution is broader, which is more in accordance with the free radical mechanism of the copolymerization. It can be speculated that the two very high molecular weight distributions arise from the formation of aggregates when the copolymer is in solution. In order to determine whether aggregations occur at these extremely high molecular weights, fractions were taken at the different peak elution times in order to investigate the chemical composition of each fraction. The different fractions were analysed by FT-IR. Fractions of both low and intermediate molecular weights displayed similar spectra to that of the dry copolymer sample. On the other hand, the fraction with the very high molecular weight distribution exhibited an absorption band at 1659 cm^{-1}, indicating the presence of a carbon-carbon double bond. This shows that a residual amount of MET monomer forms micelles or aggregates with the copolymer, giving rise to the presence of the high molecular weights distributions. Moreover, the intermediate molecular weight probably arises from aggregation of copolymers in THF.

Table 5.   Molecular weight of MET-PS and MET-MMA copolymers, determined by GPC (standard PS)

Copolymer	M_n (g/mol)	M_w (g/mol)	M_w/M_n
MET-1.5-PMMA	383 000	543 000	1.42
	215 000	368 000	1.71
	160 000	528 000	3.29
MET-3-PMMA	758 000	950 000	1.25
	481 000	793 000	1.65
	203 000	515 000	2.53
MET-6-PMMA	574 000	204 000	3.56
	132 000	403 000	3.05
MET-1.5-PS	407 000	689 000	1.69
	893 000	205 000	2.30
	332 000	108 000	3.24
MET-3-PS	612 000	105 000	1.71
	244 000	381 000	1.56
	473 000	133 000	2.82
MET-7-PS	678 000	965 000	1.42
	102 000	294 000	2.89
	394 000	919 000	2.33

The thermal stability of the copolymers was determined by TGA. All the samples showed a multi-step weight loss. The first weight loss was gradual, with an onset at about 100 °C (which is probably due to water bound to the sodium sulfonate groups of MET). The onset of the second weight loss appeared at about 300 to 400 °C, which shows up the onset of the real degradation of the copolymers. As expected for the MET-PS copolymers, the residual weight of the samples after being heated at 700 °C increases as the MET ratio increases, because of the higher content of sulphur and sodium. Nevertheless, one should note that the residual weight observed for the MET-PMMA copolymers were unchanged with the ratio of MET.

The thermal behaviour of the copolymers, as studied by DSC, showed a weak melting transition on the heating curves, while no T_g could be detected. Due to the weakness of the energies of the transitions, we have reported only the maximum temperature in Table 6 (including the crystallization temperatures observed while cooling the samples). Because PMMA and PS are amorphous, both melting and crystallization temperatures probably come from some organization of the MET monomer units. Due to the strong thermodynamic incompatibility between the PS matrix and the ionic polar head of the polymerized MET, it is believed that the transitions (melting and crystallization) observed originate from associations of the ionic heads of the copolymerized MET surfmers. Such behaviour has been previously reported for similar copolymers containing a carboxylate polar head with a divalent counter-cation M^{2+} [32].

Table 6. Temperatures of melting (T_m) and crystallization (T_c) of MET-PS and MET-MMA copolymers, determined by DSC

Copolymer	T_m (°C)	T_c (°C)
MET-1.5-PS	174.0	125.6
MET-3-PS	141.9	119.6
MET-7-PS	144.7	118.7
MET-1.5-PMMA	148.3	122.5
MET-3-PMMA	144.9	125.4
MET-6-PMMA	138.3	102.5

TEM of a 50 nm microtomed sample of MET-7-PS is shown in Figure 6. A lamellar nano-structure is formed, with spacing between the lamellae around 3 nm. Note that the expected nano-structure thickness should be around 3.8 to 4 nm if the lateral chains of the MET units are considered stretched (which is unlikely). Moreover, from the texture of the TEM picture, the organization of the surfmer repeat units could be considered as bead-like structures, so that the order is not perfectly lamellar but incorporates spheroid structures as well. Due to the thermodynamic incompatibility between the polar heads of the MET and the PS matrix, and taking into account the structure observed by TEM, one can speculate that the MET repeat units self assemble via the formation of intermolecular bridges (see Figure 7).

Figure 6.   TEM pictures of MET-7-PS

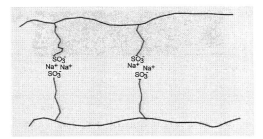

Figure 7.   Interchain interactions between the ionic heads of the MET monomer
            units

Conclusion

The monomer MET was synthesized an its thermal behaviour studied by TGA, DSC,
and microscopy under cross-polarized light. It presents a thermotropic behaviour, with
a mesophase stable from 88 °C up to 107 °C.

Pursuing the objective to study the organization of MET in a non-polar amorphous
polymer matrix, two series of copolymers that contained 1.5 to 6 mol% of MET units
and S or MMA were successfully synthesized. The resulting copolymers displayed a
weak thermal transition that can be assigned to the organization of the side chain of
the MET monomer units in the amorphous polymer matrix. Furthermore, and despite
of the low MET content, MET monomer units self assemble in pseudo-lamellar

structures. This is occurring because of the incompatibility between the MET sulfonate groups and the non-polar polymer matrix.

Acknowledgments

The support of the National Research Foundation of South-Africa (NRF) and the Deutsches Kunststoff-Institut (DKI) are gratefully acknowledged.

[1]   S. M. Hamid, D. C. Sherrington, *Polymer* **1987**, *28*, 325-331.
[2]   S. M. Hamid, D. C. Sherrington, *Polymer* **1987**, *28*, 332-339.
[3]   M. Summers, J. Eastoe, *Adv. Coll. Int. Sci.* **2003**, *100-102*, 137-152.
[4]   A. Laschewsky, *Adv. Polym. Sci.*, **1995**, *124 (Polysoaps/Stabilizers/Nitrogen-15 NMR)*, 1-86.
[5]   K. Holmberg, *Prog. Org. Coat.*, **1992**, *20(3-4)*, 235-237.
[6]   K. Nagai, *Trends Polym. Sci.*, **1996**, *4(4)*, 122-127.
[7]   K. Tauer, *Polymeric Dispersions: Principles and Applications*, ed. Series ed., Kluwer Academic Publishers, Dordrecht 1997.
[8]   A. Guyot, in: *Surfactant Science Series*, K. Holmberg, Ed., Marcel Dekker, New York 1998, p. 301-332.
[9]   J. M. Asua, H. A. S. Schoonbrood, *Acta Polymerica* **1998**, *49(12)*, 671-686.
[10]   C. Larpent, E. Bernard, J. Richard, S. Vaslin, *Macromolecules* **1997**, *30*, 354.
[11]   B. Tieke, M. Dreja, *Macromol. Rapid. Comm.* **1996**, *17*, 825.
[12]   B. Tieke, M. Dreja, W. Pyckhout-Hintzen, *Macromolecules* **1998**, *31*, 272.
[13]   B. Tieke, M. Pyrasch, *Coll. Polym. Sci.* **2000**, *278*, 375.
[14]   M. P. Pileni, A. Hammouda, T. Gulik, *Langmuir* **1995**, *11*, 3656.
[15]   R. A. Mackay, M. P. Pileni, N. Moumen, *Coll. Surf. A* **1999**, *151*, 409.
[16]   L. Tichagwa, C. Gotz, M. Tonge, R. D. Sanderson, H. Pasch, *Macromol. Symp.* **2003**, *193*, 251-260.
[17]   H. A. S. Schoonbrood, M. J. Unzue, O.-J. Beck, J. M. Asua, A. M. Goni, D. C. Sherrington, *Macromolecules* **1997**, *30*, 6024-6033.
[18]   O. Soula, A. Guyot, N. Williams, J. Grade, T. Blease, *J. Polym. Sci., Part A: Polym. Chem.* **1999**, *37*, 4205-4217.
[19]   S. Roy, P. Favresse, A. Laschewsky, J. C. de la Cal, J. M. Asua, *Macromolecules* **1999**, *32*(18), 5967-5969.
[20]   P. Favresse, A. Laschewsky, C. Emmermann, L. Gros, A. Linsner, *Eur. Polym. J.* **2001**, *37*, 877-885.
[21]   P. Favresse, A. Laschewsky, *Polymer* **2001**, *42*, 2755-2766.
[22]   M. J. Unzué, H. A. S. Schoonbrood, J. M. Asua, O.-J. Beck, A. M. Goni, D. C. Sherrington, K. Stahler, K.-H. Goebel, K. Tauer, M. Sjoberg, K. Holmberg, *J. Appl. Polym. Sci.* **1997**, *66*, 1803-1820.
[23]   A. Xenopoulos, J. Cheng, M. Yasuniwa, B. Wunderlich, *Molecular Crystals and Liquid Crystals Science and Technology, Section A: Molecular Crystals and Liquid Crystals*, **1992**, *214*, 63-79.
[24]   B. C. Wunderlich, Wei, *ACS Symposium Series*, **1996**, *632 (Liquid-Crystalline Polymer Systems)*, 232-248.
[25]   A. T. Guyot, *Adv. Polym. Sci.*, **1994**, *111(Polymer Synthesis)*, 43-65.
[26]   A. Guyot, *Current Opinion in Colloid & Interface Science*, **1996**, *1*(5), 580-586.
[27]   A. Guyot, K. Tauer, J. M. Asua, S. Van Es, C. Gauthier, A. C. Hellgren, D. C. Sherrington, A. Montoya-Goni, M. Sjoberg, O. Sindt, F. Vidal, M. Unzue, H. Schoonbrood, E. Shipper, P. Lacroix-Desmazes, *Acta Polymerica* **1999**, *50*(2-3), 57-66.
[28]   T. Otsu, T. Ito, M. Imoto, *Kogyo Kagatu Zashi* **1966**, *69*, 986.
[29]   A. Urretabizkaia, J. M. Asua, *J. Polym. Sci., Polym. Chem.* **1994**, *32*, 1761.
[30]   S. K. Sinha, A. I. Medalia, *J. Am. Chem. Soc.* **1957**, *79*, 281.
[31]   H. H. Freedman, J. P. Mason, A. I. Medalia, *J. Org. Chem.* **1958**, *23*, 76.
[32]   R. Sauerwein. 1997, Technische Hochschule: Darmstadt.

Macromol. Symp. **2004**, *214*, 231-240

Strain-Controlled Tensile Deformation Behavior and Relaxation Properties of Isotactic Poly(1-butene) and Its Ethylene Copolymers

Mahmoud Al-Hussein, Gert Strobl*

Physikalisches Institut, Albert-Ludwigs-Universität, Hermann-Herder-Str. 3, 79104 Freiburg, Germany

E-mail: gert.strobl@physik.uni-freiburg.de

Summary: The tensile deformation behaviour of poly(1-butene) and two of its ethylene copolymers was studied at room temperature. This was done by investigating true stress-strain curves at constant strain rates, elastic recovery and stress relaxation properties and in-situ WAXS patterns during the deformation process. As for a series of semicrystalline polymers in previous studies, a strain-controlled deformation behaviour was found. The differential compliance, the recovery properties and the stress relaxation curves changed simultaneously at well-defined points. The strains at which these points occurred along the true stress-strain remained constant for the different samples despite their different percentage crystallinities. The well-defined way in which the different samples respond to external stresses complies with the granular substructure of the crystalline lamellae in a semicrystalline polymer.

Keywords: deformation; poly(1-butene); relaxation; recovery; yielding

Introduction

In previous papers we reported on the tensile deformation behavior of several polyethylenes and s-polypropylenes based upon measurements of true stress-strain curves, elastic-recovery properties, and texture changes at different stages of the deformation process. The results [1-3] showed that there is a general scheme that governs the behavior. Along the true stress-strain curve, the differential compliance, the recovery properties, and the crystallite texture change simultaneously at well-defined points. The strains at these points are invariant over various crystallinities, strain rates and drawing temperatures. In contrast to this, the corresponding stresses vary considerably.

© 2004 International Union of Pure and Applied Chemistry   DOI: 10.1002/masy.200451016

In a subsequent paper, we reported on the tensile deformation of a set of poly(1-butene)-based samples with different crystallinities. Poly(1-butene) (P1B) is renowned for its good creep resistance and the retaining of its mechanical properties at elevated temperatures. The mechanical tests were now further expanded, by the inclusion of stress relaxation measurements.

Experimental

Sample characteristics and preparation

Three different P1B-based samples (provided by Basell, Louvain-La-Neuve, Belgium), were used in this study. Their characteristics are shown in Table 1.

Table 1. Sample Characteristics of P1B

sample	grade	ethylene %	$T_m/°C$	crystallinity %
A	PB0300	0	123	47
B	PB8220	6	119	37
C	PB010	>6	108	16

Compression-moulded sheets were prepared from the different samples. After cooling, the sheets were stored at room temperature for at least 40 days to allow the samples to transform into their stable form I. Then dog-bone-shaped specimens of 6.5 x 4 mm were cut from the sheets for testing.

True stress-strain curves

True stress-strain curves at a constant Hencky strain rate were obtained using a video-controlled tensile testing apparatus. It employs a video camera connected to a computer to control the deformation by regulating the cross-head speed in a way that keeps the Hencky strain rate at a constant value.

Recovery and stress relaxation

The same apparatus was also used to investigate the elastic recovery properties at different stages of the deformation process. This was done by carrying out a step-cycle test. A specimen is stretched first to a predetermined strain, ε_{tot}, and then it is brought back to a zero stress. The remaining strain at this point represents the plastic part, ε_{pla}, and the difference between the total and plastic strain gives the elastic part, ε_{ela}.

The apparatus was also used for the stress relaxation measurements. A test specimen is stretched first to a predetermined strain at a constant strain rate, as explained above. Subsequently, the strain is held constant and the force decay is followed with time. The prior stretching was performed at a constant true strain rate of 0.005 s^{-1} for all measurements.

Results

Deformation Behavior

Figure 1 shows true stress-strain curves obtained at a constant strain rate of 0.005 s^{-1} at room temperature for the different samples. None of the samples showed necking down during deformation despite their different crystallinities. The different curves are similar to each other, resembling a rubber-like behavior.

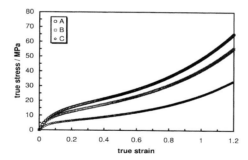

Figure 1.   True stress-strain curves obtained at true strain rate of 0.005 s^{-1} of samples A, B and C.

Step-cycle tests were used to decompose the imposed strain into elastic and plastic parts that were recoverable and irrecoverable, respectively, on the time scale of the experiment and at room temperature. A representative example of a step-cycle test is given in Figure 2. The total, plastic, and elastic strain was then plotted as a function of the true stress (Figure 3).

As can be seen, at the beginning of the deformation process, both the plastic and elastic strain increased with increasing total strain. This carried on until the total strain reached a certain value, at which the elastic strain assumed a plateau value, and any further increase in the total strain proceeded by only increasing the plastic strain. This occurs at a point, which we refer to as point C. The interesting feature is that the total strain at point C was common for the different samples, $e_H \approx 0.7$.

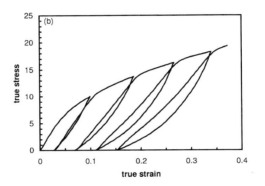

Figure 2.   A representative example of the step-cycle test (sample A).

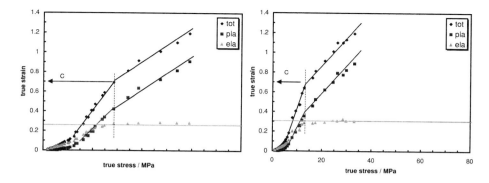

Figure 3.   True total and elastic strain at different imposed true stresses for the different samples, A (left) and C (right).

Looking more closely at the low deformation region revealed another two transition points, A and B, which occurred again at the same strains of 0.05 and 0.1, respectively, for all samples. Figure 4 shows typical examples of stress relaxation tests performed at different strains. It can be seen that the stress always decreased and eventually approached a final plateau value. There was no substantial change in the shape of the relaxation curve up to a strain of 0.7. After this strain the amount of stress relaxation increased.

To learn more about a possible cause of this increase we examined the strain rate effect. Figure 5 shows the stress relaxation curves obtained after stretching to a true strain of 1, at three different strain rates. As seen, whilst the plateau value changed only slightly, the initial stress drop showed a strong dependence on the strain rate. Obviously this drop is associated with the cessation of an instantaneous viscous flow component.

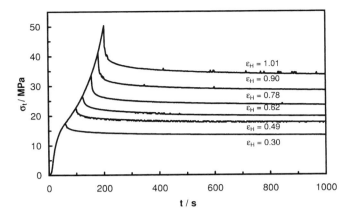

Figure 4.   Stress relaxation test at the indicated Hencky strains (sample A).

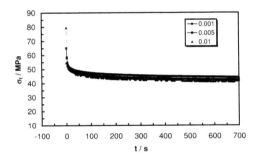

Figure 5.   Stress relaxation curves after deformations with the indicated Hencky strain rates to $\varepsilon_H = 1$ (sample A).

It seems that once the deformation stops, this instantaneous flow ceases, and the extra force that was required to keep this flow vanishes. It is evident from Figure 4 that different curves relaxed to different plateau values, which we will refer to as the unrelaxed stress, σ_{unr}. Also notable is the fact that none of the curves relaxed to zero stress even at high strains. In order to quantify the relaxed stress values, we define σ_{rel} as the difference between σ_0 and σ_{unr}. Figure 6 shows the variations of σ_0, σ_{unr} and σ_{rel} with strain, together with a constant strain rate response. As it can be seen, they all increased with increasing strain.

To investigate whether the relaxation has any effect on the recovery properties we again performed step-cycle tests, but now at the end of the stress relaxation test. Results are shown in Figure 7. The consequence of the stress relaxation was a minor loss in the elastic strain. The position of point C remains unchanged, as shown by Figure 8.

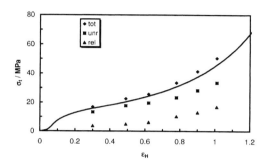

Figure 6. A true stress-strain curve (solid line) together with total, unrelaxed and relaxed stresses at different imposed Hencky strains (sample A).

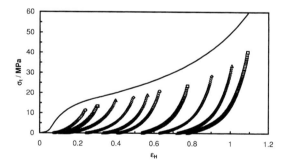

Figure 7. Stress relaxation followed by step-cycle tests at different Hencky strain values (sample A).

Discussion

Results of the experiments gave us the opportunity to assess the influence of the crystallinity on the deformation behavior for a set of P1B-based samples. As expected, increasing the volume fraction of the crystallites increases the internal friction. This results from inter-and intra-lamellar shear processes. In addition, the different samples showed a common rubber-like deformation behavior. As the deformation proceeded, both the differential compliance and the elastic properties changed at three transition points, A, B, and C, at increasing strain. A Hookean elasticity was exhibited first until point A, where a plastic strain started to occur. Simultaneously, the differential compliance showed a slight increase. This continued up to point B, where another more pronounced increase in the differential compliance took place. Finally, at point C the elastic strain reached a plateau value, and the differential compliance decreased this time. WAXS experiments showed that at point C sharp spots, typical of a fibrillar structure, appeared on the equator.

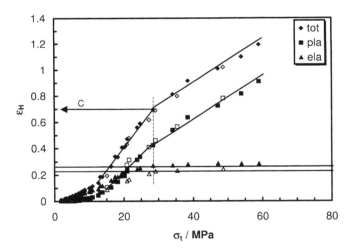

Figure 8. Filled symbols are the total, plastic and elastic strains at different imposed true stresses, as obtained from Figure 7, the open symbols are the same strains as obtained from step-cycle tests without a prior relaxation (Figure 3, left).

Changing the crystallinity of the different samples had no effect on the strain values at which the transition points occurred along a true stress-strain curve. The fact that the strain values of the transition points are unaffected by changing the crystallinity or other factors demonstrates the capability of semicrystalline polymers to accommodate any imposed strain in a rather well defined way. This can only be achieved if there are a sufficient number of internal degrees of freedom available for each semicrystalline polymer by virtue of its structure. This is consistent with recent studies of semicrystalline polymers showing a granular substructure of the crystal lamellae [4]. The chain slip processes at the boundary region between the adjacent blocks in a single lamella then provide the extra degrees of freedom needed.

A semicrystalline polymer can be viewed as two interpenetrating networks, a crystalline one intermingled with an entangled amorphous one. Consequently, both of these networks will contribute to any imposed deformation. The extent of each contribution varies during the course of deformation. At low strains, the crystalline network contribution dominates through changing the coupling and invoking some coarse slips of the crystalline blocks (point B). As the strain is increased, the entangled amorphous region becomes increasingly strained. This carries on until a critical strain is reached where the force generated from the entangled fluid regions reaches a critical value that would be able to destroy the crystallites (point C). At this stage the dominant mechanism becomes the disaggregation of the crystallites and apparent recrystallization into fibrill.

Our results also indicate a direct relation between the deformation mechanisms that are active at a certain strain and the subsequent relaxation behaviour. This is demonstrated first by the qualitative change in the shape of the stress relaxation behaviour at point C where an additional viscous force sets in. At low strains and up to the point C, the deformed sample responds mainly by interlamellar shearing and slip processes at the interfaces of the crystalline blocks. Therefore, in order to realize the imposed strain, the blocks have to undergo a continuous readjustment. During this process the majority of the blocks remains intact, but they are away from their equilibrium position. If the crosshead is stopped at any strain in this stage, the blocks pass from their current non-equilibrium states to a new equilibrium state by local movements. This relieves the stress locally and leads to a decrease in the free energy. At point C the initial blocks start to disintegrate and break away from each other, resulting in a new crystal skeleton with properties different from the original one. The disintegration process contributes to the viscous force and, correspondingly, shows up in the initial decrease in the stress relaxation curves.

[1]   R. Hiss, S. Hobeika, C. Lynn and G. Strobl. *Macromolecules* **1998**, *32*, 4390.

[2]   S. Hobeika, Y. Men and G. Strobl. *Macromolecules* **2000**, *33*, 1827.

[3]   Y. Men and G. Strobl. *J. Macromol. Sci.* **2001**, *40*, 775.

[4]   G. Strobl. *Eur. Phys. J. E* **2001**, *3*, 165.

Technology and Stress Relaxation of Biaxially Oriented Polyolefin Shrink Films

Arthur Bobovitch,[1,2] *Yakov Unigovski,*[2] *Albert Jarashneli,*[2] *Emmanuel M. Gutman*[2]

[1] Syfan Saad (99) Ltd., Israel

[2] Dept. of Materials Engineering, Ben-Gurion University of the Negev, Israel

Summary: In the present study the influence of the heating rate on the stability of the double bubble technological process was investigated. It was shown that increasing the heating rate decreases the stability of the process and causes the lower elongation of the films produced. The morphological transformations of linear low density polyethylene (LLDPE) film were explained using X-ray and transmission electrom microscopy (TEM) methods. The stress relaxation behavior of co-extruded LLDPE/ethylene-vinyl-acetate (EVA) film was studied using the relaxation time spectrum approach. The influence of vinyl-acetate (VA) content in EVA copolymers on the relaxation time spectrum was observed.

Keywords: morphology; polyolefin film; relaxation time spectrum; stress relaxation

Introduction

Polyethylene (PE) films are widely used in the packaging industry. These films are extensively used in flexible packaging of a wide spectrum of products. They provide a very good combination of physical and mechanical properties, which meet the demands and technical requirements of the packaging industry. The first films produced by "double-bubble" technology were made from PVC, but today, the trend is to leave PVC for polyolefins like polyethylene, polypropylene, ethylene-vinyl-acetate. The main reasons for this trend are: toxicity of the degradation products of PVC (mainly HCl) and lack of the approval for plasticisers, that are an important part of PVC based formulations, to be in contact with food.

The most widely used packaging technique associated with PE films is the shrinkage method. In this method the film wraps the packaged item, producing a bag. This bag is introduced into a heated tunnel. If the film is oriented it shrinks and wraps the item smoothly. There are several processes used for the production of these oriented films. One of the best known is biaxial orientation by the "double-bubble" process or tubular orientation process. In this process the primary extruded tube is quenched, reheated to a temperature below the melting point, and then oriented in both machine direction (MD) and transverse direction (TD) simultaneously. The advantage of this technique is the resulting balanced properties of the film in both directions[1]. Biaxially oriented films possess exceptional clarity, superior tensile properties, improved flexibility and toughness, improved barrier properties and the unique property of engineering shrinkability[2].

The production lines for the double bubble process have changed over the past fifteen years. The first production lines were equipped with water stream heating ovens, but recently IR ovens have been introduced. These ovens increase the heating capacity of the lines but cause problems with the stability of the second bubble. The present work describes a way in which the stability of the second bubble can be increased.

Despite its technological advantages, the double-bubble process has not been studied in depth from the viewpoint of morphological transformations taking place with the polymer during production. One of the goals of this study is to describe the transformations taking place within the polymer at different stages of the technological process. An understanding of this behavior should lead to improvement of technological parameters as well as to better materials choice during construction of film formulations.

During different production stages (slitting, folding etc.) the film is stressed, usually below the yield point. The behavior of the stressed film inside the rolls can be characterized by stress-relaxation. The film is under stress, it is deformed but cannot relax its deformation, which is constant. Hence, a study of the stress-relaxation process, which is usually characterized by relaxation time spectra, is of interest from both scientific and technological viewpoints.

Relaxation time spectra are of fundamental interest.

$$H(\tau) = -[dE(t)/d\ln t]_{t=\tau} \qquad\qquad (1)$$

where H (τ) is the relaxation time spectrum and E(t) is the relaxation modulus. It is often used to discuss the molecular theory of materials and their morphology [3]. Some authors have investigated the relaxation time spectra of crosslinked materials. Fedors et al [4] investigated the stress-relaxation behavior of (EVA) cross-linked to different gel contents. They tried to explain long-term relaxation behavior using the stress-relaxation tests at different temperatures, but no conclusions about the influence of crosslinking density on morphology and relaxation behavior of the polymer were reached. Gotlib and co-authors [5-7] explained relaxation processes in crosslinked polymer networks. They investigated the influence of interchain motion and friction on the relaxation process. The models of segment relaxation were also studied. The morphology of crosslinked polymers was however not discussed. It is not clear what the influence is of varying degrees of crosslinking on the relaxation time spectrum.

In our previous work we investigated the morphology and stress relaxation of composite films with polypropylene in the outer layers and with LLDPE in the skin layer. The morphology and stress relaxation behavior of crosslinked LLDPE films was also studied [8-10]. In the present paper we will discuss the relaxation behavior of the film coextruded with LLDPE and EVA. These films are often used in paper products packaging (journals, envelopes, etc.) In these applications lower shrinkage forces are required in order not to destroy the packed item.

Experimental

In order to investigate the influence of the heating rate on the stretching process and on the properties of the final films a sheet from LLDPE (density 0.920g/cm^3 and melt flow index (MFI) 1) with average thickness of 375 microns was extruded. The films were produced by stretching the extruded sheets at a KARO IV laboratory stretcher produced by Brueckner, Germany. The films were stretched simultaneously in MD and TD. The orientation ratio was 1:5 in each direction. The different heating rates were applied. The mechanical properties of the films with the thickness of 15 microns, achieved after biaxial stretching, were tested according to ASTM D 1894 using a LLOYDS LRX tensile tester. Differential scanning calorimetry (DSC) at different heating rates was used to determine the influence of heating rate on polymer morphology.

X-ray diffraction was used to investigate the morphological transformation in LLDPE during the double bubble process. The X-ray diffraction patterns of the extruded sheet of 0.375 mm polyethylene were recorded at different temperatures with a Philips X-ray diffractometer (type PW-1130) using Co K_α radiation and a Fe – filter at the angular range $17 - 40°$, scanning rate $2°$/min, an operating voltage of 40 kV and a current of 30 mA. In order to carry out the X-ray experiments at elevated temperatures a special unit was designed (Figure 1).

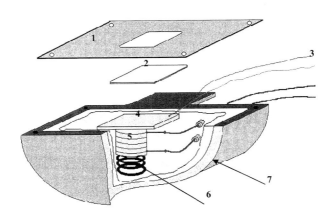

Figure 1.   Specimen's holder specially designed to carry out X-ray experiments at elevated temperatures: 1 – cover, 2 – specimen, 3 – thermocouple, 4 – ceramic platform, 5-heater, 6 – spring, 7 – isolation.

Samples for transmission electron microscopy were first stained in RuO_4 for several days. Ultrathin sections (60 nm) were prepared at room temperature, using a diamond knife with a LEICA Ultramicrotome. TESLA BS 500 TEM was operated.

In order to investigate the stress-relaxation behavior a five-layer co-extruded crosslinked film was used. The structure of the film was LLDPE/EVA/LLDPE/EVA/LLDPE. The thickness of each EVA layer was 20 %, the thickness of LLDPE skin layers was 10 % and the thickness of LLDPE core was 40 %. The total film thickness was 15 microns. The EVA copolymers with VA contents

4 %, 12 % and 19 % were used.

For the stress relaxation measurements, samples with 6 mm width were carefully cut from the film. The experiments were carried out with Zwick 1405 tensile tester with a 10 N load cell.

Results and Discussion

Morphology transformations during the double-bubble process

The morphology transformation of any semicrystalline polymer during the double-bubble process can be presented by Figure 2.

After extrusion the melt is quenched with water, preventing spherulitic structure formation. Lamellae with different thicknesses can be observed (Figure 3). The quenched polymer is then reheated to below the melting point. The crystallinity of the polymer decreases. Figure 4 shows that, after heating to 80 °C, the polyethylene peak in the X-ray diffraction pattern decreases dramatically.

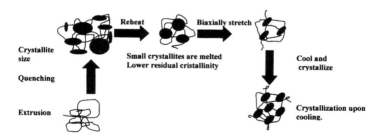

Figure 2.   Morphology transformations in a semicrystalline polymer during the double-bubble process

Figure 3.   TEM photograph of the quenched LLDPE (X 120,000)

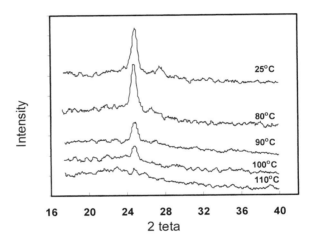

Figure 4.   X-ray diffraction patterns of LLDPE as a function of temperature

After reheating to temperatures below the melting point the polymer is stretched in both MD and TD simultaneously, with the same orientation ratio. The lamellae structure transforms into fibrilious-like with the fibrila oriented at about 45 º between MD and TD.

Figure 5.   TEM photograph of biaxially oriented LLDPE film (X 120,000).

Influence of the heating rate on the process and on the film properties

As has been mentioned, the first production lines for the double-bubble process were equipped with a hot water steam-based oven. As technology developed, these ovens were later substituted by IR-based ovens. The main reason for this was to provide quicker heating and increased output. The heating rate in an IR-oven is about three times higher than the heating rate in a water/steam oven. It was however observed that with the IR-oven the bubble stability was decreased and the film became less elastic. In order to stretch LLDPE biaxially the polymer has to be heated to a specific orientation temperature. DSC measurements showed that with an increasing heating rate the melting curve of the polymer is sharper [11]. This means that the window of orientation temperatures becomes narrower with increasing heating rate. Indeed, when the heating rate is high, a small temperature change will cause significant change in the polymer crystallinity.

In order to understand the influence of the heating rate on the stretchability of LLDPE, ten samples were stretched biaxially at different heating rates and the number of broken samples was calculated. When the heating rate reached 12 °C/min the number of unsuccesfull tests was more than 60 %. Any further increase in the heating rate resulted in it being impossible for LLDPE to be stretched; 100 % of samples were broken after stretching.

The influence of the heating rate on mechanical properties of a LLDPE film is shown in Figure 6. With an increase in heating rate the ultimate tensile stress increases and elongation decreases dramatically.

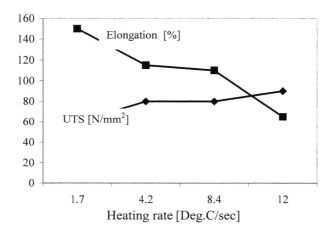

Figure 6. An influence of heating rate on ultimate tensile stress (UTS) and elongation of LLDPE-film

Stress-relaxation of co-extruded LLDPE/EVA film

Relaxation time spectra of the LLDPE/EVA co-extruded films, containing EVA with different VA content, are shown in Figure 7.

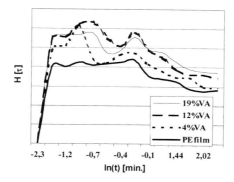

Figure 7.   Relaxation time spectra of LLDPE/EVA co-extruded films

Two main peaks can be observed in each spectrum. At *ln(t)* = – 0.3 the peak corresponds to the crystalline phase of the polymers and the peak at about *ln(t)* ≈ – 0.8 corresponds to the crosslinked amoprphous phase[9]. One can see that the difference between the relaxation of the amorphous and of crystalline phases clearer when EVA copolymers are included in the formulation. Higher peak intensity in the relaxation time spectra of co-extruded LLDPE/EVA, in comparison with the spectra of LLDPE film, means that the observed decrease in stress was more significant in the co-extruded film than in the LLDPE film. Hence, the remaining stresses in co-extruded film will be lower that in LLDPE film.

Conclusions

The morphology transformation of LLDPE during the double-bubble process was investigated and explained. It was shown that the morphology of the oriented film is fibril-like, with the fibril orientation at 45 degrees between MD and TD. It was also shown that the heating rate influences the stability of biaxial orientation as well as the properties of the film; the higher the heating rate the worse the process stability and the lower the elongation of the films produced.

The study of relaxation time spectra of crosslinked LLDPE film and crosslinked coextruded LLDPE/EVA film shows two main peaks, corresponding to crystalline and to crosslinked amorphous phases. The difference in relaxation of these two phases is clearer in coextruded films. Practically, that means that the coextruded LLDPE/EVA films have to be suspended between different technological stages (extrusion, slitting, centerfolding) for the same time periods as LLDPE film in order to relax internal stresses. The observed relaxation behavior of coextruded LLDPE/EVA film (higher peaks on relaxation time spectra) can provide the worse performance of this film during its winding (wrinkling during suspension time). It is for this reason that it is recommended that the tension during winding of this co-extruded film be decreased.

[1]   K. R. Osborn, K. R. Jenkins, in: *"Plastic Films"*, Technomic Publishing Co., Lancaster, PN 1992.
[2]   C. J. Benning, in: *"Plastic Films For Packaging"*, Technomic Publishing Company, Lancaster, PN 1983.
[3]   C. Friedrich, H. Braun, J. Weese, *Polym. Eng. Sci.* **1995**, *35*, 1661.
[4]   R. F. Fedors, S. Y. Chung, S. D. Hong, *J. Appl. Polym. Sci.* **1985**, *30*, 2551.
[5]   Yu Ya Gotlib, *Pure & Appl. Chem.* **1981**, *53*, 1531.
[6]   Yu Gotlib, G. Golovachev, *Journal of Non-Crystalline Solids* **1994**, *172*, 850.
[7]   A. A. Gurtovenko, Yu Ya Gotlib, *Macromolecules* **1998**, *31*, 5756.
[8]   A. Bobovitch, E. M. Gutman, S. Henning, G. H. Michler, *J. Applied Polymer Sci.* (accepted).
[9]   A. Bobovitch, E. M. Gutman, S. Henning, G. H. Michler, *Materials Letters* **2003**, *57/16-17*, 2579.
[10] A. Bobovitch, E. M. Gutman, S. Henning, G. H. Michler, Y. Nir, *J. of Plastic Film & Sheeting* (in press).
[11] V. B. F. Mathot, in: *"Calorimetry and Thermal Analysis of Polymers"*, Hanser Publisher, Munich, Vienna, New York 1994.

Macromol. Symp. **2004**, *214*, 251-259

Failure Envelope Curves in Polyethylene Solids

Koh-hei Nitta, * *Takashi Ishiburo*

Department of Chemistry and Chemical Engineering, Kanazawa University, 2-40-20 Kadatsuno, Kanazawa, 920-8667 Japan
E-mail: nitta@t.kanazawa-u.ac.jp

Summary: The present work demonstrates that the failure envelope analysis can be applied for characterizing the ultimate tensile properties of polyethylene solids in which the inhomogeneous necking process is avoided. As a result, the ultimate properties are essentially identical to those of vulcanized rubbers above the glass transition temperature, suggesting that tie molecules connecting the fragmented lamellar clusters transmit the external led to the fracture site in the same manner as cross-linked rubbers do. Consideration of this crystal-network model may provide information about the molecular processes that lead to rupture. Furthermore, the present analytical method can possibly be developed for predicting rupture times when different types of tests, such as constant drawn, constant stress and constant rate of stress, are conducted.

Keywords: failure envelope curve; polyethylene; ultimate strain; ultimate stress

Introduction

Polyethylene is one of the most widely used polymers because it is chemically simple and it has a wide applicability in practical use. However, polyethylene solids exhibit complicated tensile behavior accompanying inhomogeneous deformation, such as neck propagation and plastic flow, compared to rubber-like materials. Such complicated deformation processes make it difficult to accurately characterize the ultimate tensile properties as well as the deformation behavior. The inhomogeneous deformation due to the neck propagation often causes strong dependence of the ultimate parameters, such as the elongation at break or the time to break, on the shape and length of the test piece. This is because the neck gradually propagates through the sample with increasing time of draw, with the result that the elongation at break or the time to break increases with increasing length of the test piece, because a specimen is elongated at a constant elongation rate. In this work, by eliminating such overestimation of the ultimate elongation due to the neck propagation process, the ultimate properties of polyethylene solids can be adequately investigated.

DOI: 10.1002/masy.200451018

Experimental [1]

The polymers used in this work were metallocene-catalyzed polyethylenes: a high-density polyethylene (HDPE) and a linear low density polyethylene (LLDPE). Their molecular characteristics are listed in Table I. Polymer pellets were melted in a mould between flat metal plates in a hot press at 190 °C. A pressure of 10 MPa was applied to remove air bubbles and excess melt and then the samples quenched at 0 °C were prepared for the measurements.

Double edge-notched specimens with a gauge length of 2 mm, a ligament length of 2 mm, a width of 8 mm and overall length of 40 mm were used for the tensile tests. They were punched out from the compression-moulded sheets of 200 μm thickness. The shape of the tensile specimen was designed in such a way as to eliminate/avoid the plateau region or necking region in the stress-strain curves. In the specimen, the neck immediately reaches a whole portion of the gage length just after the neck is initiated, with the result in that uniform deformation can be realized throughout the tensile test.

The sample specimens were deformed using a small tensile tester designed and built in our laboratory. It was equipped with a temperature-controlled environmental chamber. The samples were elongated over a wide range of elongation rates and temperatures.

Table 1. Molecular characteristics of polyethylene samples

Sample	$Mw \times 10^{-4}$ g/mol	$Mn \times 10^{-4}$ g/mol	Mw / Mn	SCB mol% [1]
HDPE	7.3	2.9	2.5	0
LLDPE	7.8	4.4	1.8	0.88

[1] short chain branching

Results and Discussion

Figure 1 shows a series of stress-strain curves for HDPE measured over a wide range of temperatures from –40 ° to 100 °C at 5 mm/min, and a wide range of elongation rates from 0.125

to 5 mm/min at 25 °C. At below 0 °C, the results show a strong strain-hardening effect immediately after the yield peak. The strain-hardening effect is reduced as the test temperature increases. As seen in Figure 1, the ultimate stress increases with decreasing temperature whereas the ultimate strain has a maximum point. Comparison between these figures reveals that the ultimate behavior measured at elevated temperatures from 25 ° to 75 °C, as shown in the top figure, is congruent with the behavior measured at reduced elongation rates from 5 to 0.125 mm/min, as shown in the bottom figure. Thus, the decrease in elongation rate has a similar influence to an increase in temperature. This experimental fact will be explained on the basis of the time/temperature superposition principles later.

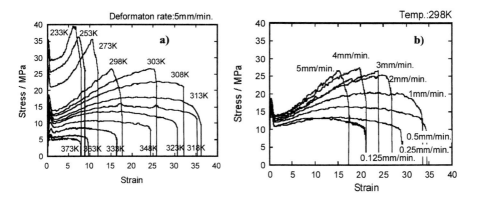

Figure 1. Temperature dependence of stress-strain curves for HDPE measured at the elongation rate of 5mm/min (a) and elongation rate dependence of stress-strain curves of HDPE measured at 25 °C (b)

Figure 2 shows the temperature and elongation rate dependences of stress-strain curves of LLDPE, respectively. As shown in these figures, the stress-strain behavior of LLDPE is greatly sensitive to the temperature compared to the deformation rate. Contrary to the case of HDPE the ultimate properties of LLDPE appear not to be based on the time/temperature superposition principles.

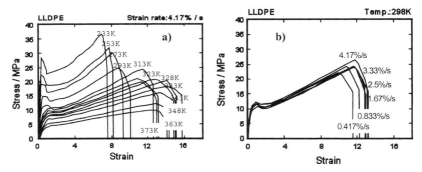

Figure 2. Temperature dependence of stress-strain curves for LLDPE measured at the elongation rate of 5 mm/min (a) and elongation rate dependence of stress-strain curves of LLDPE measured at 25 °C (b)

As is well known, elastomeric materials such as vulcanized rubber exhibit an extremely large strain to failure in uniaxial tension and they do not involve permanent plastic deformation. In the tensile deformation of these types of materials, an important experimental result has been found by Smith,[2-3] i.e. changing the deformation rate at a constant temperature has the equivalent effect on the failure strain as changing the temperature at a constant deformation rate. It is therefore very interesting to note that polyethylene solids and the elastomeric materials have the similar tendency despite the fact that they show plasticity rather than elasticity. In general, the fact that the observed values of the ultimate nominal stress σ_b and the ultimate nominal strain ε_b or the time to break t_b are interdependent is a reflection of the controlling influence exhibited by the viscoelastic response of the material. As a result of this situation a logarithmic plot of the reduced ultimate stress $\sigma_b(T_0/T)$, where T is the test temperature in Kelvin and T_0 is a reference temperature, against the strain at break ε_b, will also yield a master curve called the failure envelope. The values of σ_b are multiplied by the ratio T_0/T to correct for the temperature effect resulting from entropy elasticity. This procedure is in accordance with the kinetic theory of rubberlike elasticity, which predicts that the elastic retractive force in a specimen at a fixed extension ratio increases in direct proportion to the absolute temperature, even though the theory has no ability to treat rupture phenomena, plastic deformation nor time-dependent mechanical behavior. Here we applied this analytical procedure to the ultimate data of the present polyethylene samples, as it is.

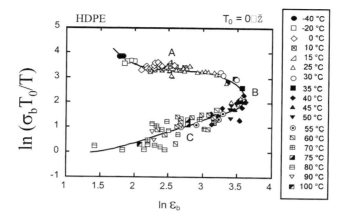

Figure 3. Failure envelope for HDPE. The data were obtained at various temperatures between -40 and 100 °C and rate of deformation at 0.5 to 5 mm/min at each temperature

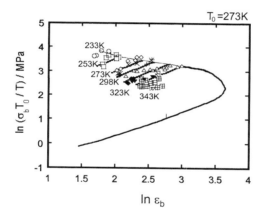

Figure 4. Failure envelope for LLDPE. The data were obtained at various temperatures between -40 and 100 °C and rate of deformation at 0.5 to 5 mm/min at each temperature

As shown in Figure 3, all the ultimate data surprisingly fall on a failure envelope curve within experimental errors for HDPE. The envelope is defined by experimental values of σ_b and ε_b, which were determined at various temperatures between –40 ° and 100 °C and at 0.5-5 mm/min elongation rates at each temperature. The reference temperature T_0 was taken to be 0 °C. It is also very interesting to note that the ultimate properties of typical plastics such as polyethylenes showing essentially energetic elasticity can be treated in the same method as those of rubber materials showing entropic elasticity.

It is likely that over an elevated temperature range above room temperature LLDPE crystals partially melt. As time-temperature superposition is normally not applicable to data obtained when partial-melting occurs, ultimate property data for LLDPE from tests under many conditions would not be expected to define a failure envelope. The ultimate data for LLDPE, which are obtained at various temperatures between −60 ° and 70 °C and at 0.5-5 mm/min elongation rates at each temperature, are shown in Figure 4. This figure indicates that those data appear to define the low-temperature segment of the failure envelope, in which the solid line of Figure 4 is the same with the line presented in Figure 3. The dotted lines show approximately the variation of the ultimate properties with strain rate over the experimental range of temperature. Along any of these lines, a point tends to move upward as the deformation rate is increased. It seems likely that if partial-melting did not occur, all data obtained between −60 ° and 70 °C would cover onto a line to provide a continuous failure envelope. The explanation for the progressive displacement of a dotted line with temperature is that each dotted line is recognized as a segment of the failure envelope of a different material, and for each material the ultimate properties are controlled by viscous effects.

The rupture point moves counterclockwise around the envelope, either as the elongation rate is increased or as the test temperature is decreased. Such a failure envelope, which is independent of time and temperature, must result, provided the well-known time-temperature superposition principle is applicable to ultimate property data determined at different temperatures and elongation rates. To determine specifically how elongation rate and temperature affect the ultimate properties, such data can be analyzed by applying the method of time-temperature superposition. The time dependence of the reduced ultimate strain at seven temperatures from 0 ° to 80 °C is exemplified in Figure 5.

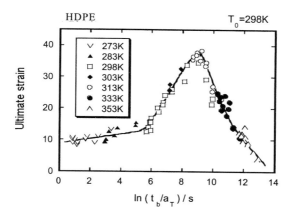

Figure 5.   The time dependence of the reduced ultimate strain for HDPE

The curves, representing data at different temperatures, were shifted along the log t axis to superimpose. Let the shift distance required to superimpose data at temperature T with those at a selected reference temperature to be a_T. To test the applicability of the Williams-Landel-Ferry (WLF) equation,[4] a plot was made of $-(T-T_0)/\ln a_T$ vs $T-T_0$, where the reference temperature T_0 was taken to be 25 °C (Figure 6).

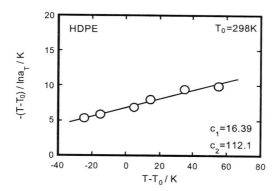

Figure 6.   WLF analysis of ultimate parameters of HDPE

As shown in Figure 6, all data between 0 ° to 80 °C fall on the straight line, indicating that the WLF equation fits the temperature dependence of ln a_T. The slope and intersection of the straight line provided the following WLF equation when the reference temperature T_0 is 25 °C.

$$\ln a_T = -\frac{16.4(T - T_0)}{112.1 + T - T_0} \tag{1}$$

In studying the linear viscoelastic properties of amorphous polymers above T_g, we have found the shift factor to be extremely useful for interconverting the effects of the temperature and experimental timescale. The conclusion drawn from the present results for PE materials was that the shift factor could also be used to interconvert the effects of the temperature and elongation rate on the ultimate properties in the same way as for rubbery elastomeric materials. This implies the importance of the noncrystalline regions in the fracture process. So far, it has been well established that crystalline polymers exhibit strong nonlinearity in their stress-strain behavior, even as small strains, with the result that not only Boltzmann superposition principles are not applicable at small strains. This agrees with the experimental fact that the stress-strain curves up to yield point, as shown in Figure 1, are very sensitive to the temperature in comparison with the deformation rate, and the time-temperature superposition cannot be applicable to the initial stress-strain regions.

Conclusion

The influence of elongation rate and temperature on the ultimate tensile properties of melt-crystallized linear PE solids was investigated using a double edge-notched specimen to avoid necking in which uniform deformation may be assumed throughout the experiment. The tensile tests were performed over a wide range of temperatures, from –40 ° to 100 °C, and of elongation rates from 0.125 to 5 mm/min. The data on ultimate properties such as tensile strength and elongation at break for different temperatures could be superimposed, by shifts along the elongation rate axis, to give a master curve as a function of the time to rupture. It was found that the ultimate stress decreases monotonously with reduced time to rupture whereas the ultimate strain has a maximum point. This behavior was the same as that of rubbery materials. The shift factors obtained from superposition of both tensile strength and ultimate strain took the form of the WLF equation, resulting in that rupture times can be predicted when the tensile tests are conducted at any experimental condition. Experimental results indicate that the ultimate properties of PE solids that show time-temperature dependence demand a physical process, which is viscoelastic in nature.

The ultimate data provided a failure envelope curve, which has the same shape as that for rubbery materials. The rupture point moves counterclockwise around the envelope either as the elongation rate is increased or the test temperature is decreased. It is very interesting to note that the ultimate properties of typical plastics such as PEs showing essentially energetic elasticity can be treated in the same method as those of rubber materials showing entropic elasticity.

[1] K. Nitta, T. Ishiburo, *J. Polym. Sci.Polym. Phys.* **2002**, *40*, 2018.
[2] T. C. Smith, *J. Polym. Sci.* **1958**, *32*, 99.
[3] T. C. Smith, *J. Polym. Sci.* **1963**, *A1*, 3597.
[4] M. L. William, R. F. Landel, J. D. Ferry, *J. Am. Chem. Soc.* **1955**, *77*, 3701.

Macromol. Symp. **2004**, *214*, 261-277

Thermal Oxidation and Its Relation to Chemiluminescence from Polyolefins and Polyamides

Lyda Matisová-Rychlá,[*1] *Jozef Rychlý,*[1] *P. Tiemblo,*[2] *J.M. Gómez-Elvira,*[2] *M. Elvira*[2]

[1] Polymer Institute of the Slovak Academy of Sciences, 842 36 Bratislava, Slovak Republic

E-mail: upolrych@savba.sk

[2] Instituto de Ciencia y Tecnología de Polímeros (CSIC) Juan de la Cierva 3, 28006 Madrid, Spain

Summary: Chemiluminescence (CL) from oxidation of polypropylene, polyethylene and polyamide has been compared and kinetic parameters based on the simplified kinetic scheme involving both bimolecular decomposition of hydroperoxides as an initiating event and correction for the oxidation spreading were determined. Induction times of oxidation determined from the autocatalytic shape of chemiluminescence intensity–time runs increase with an increasing initial molar mass of polypropylene within low molar masses up to 180 kDa regardless of the route of the polypropylene synthesis. The more complex chemiluminescence patterns in case of polyamides may be related with an increase of initially present defect structures including the terminal amino groups.

Keywords: chemiluminescence; degradation; polyamides; polyolefins; stabilization

Introduction

The oxidation of any organic material is accompanied by a very weak light emission which is obviously related to the rate of the process, however, the relation is not usually straightforward and has to be examined for any particular case. As early as in 1961 and 1964 Ashby [1] and Schard [2] pointed out that different CL intensity – time patterns exist for the series of the most frequent polymers and that there is a necessity of better understanding the CL phenomenon. Some polymers like polypropylene, polyethylene and polyamides exhibit an autocatalytic increase of the intensity of the light emission accompanying the oxidation, which is apparently coincident with autocatalytic character of the oxidation process; some give only steady decay of the light intensity

      DOI: 10.1002/masy.200451019

from an initial value. After years of a relatively active study of the CL phenomenon and its applicability in polymer oxidation some qualitative mechanistic findings may be summarized as follows [3]:

The signal intensity is determined by the quality of the polymer, by the character of its terminal groups, by the extent of previous oxidation and thermo and/or photo-oxidation history of the polymer sample expressed in concentration of hydroperoxides, carbonyl groups or other oxidized structures. The extent of oxidation corresponds to resulting mechanical properties of the polymer and to the average molar mass and its distribution. The ratio of amorphous/crystalline regions and tacticity of the polymer may affect the CL intensity – time patterns significantly.

The signal intensity is related to the sum of rates of initiating events generating free radicals, however, the contribution of respective initiating events to the total CL intensity may vary. The parallel formation of potential light emitters like excited carbonyl groups, singlet oxygen, etc. is beneficial for the increase of quantum yield of chemiluminescence. Increasing temperature and concentration of oxygen in the atmosphere surrounding the oxidized sample lead to an increase of the signal intensity. Chain breaking antioxidants and peroxide decomposers in polymers suppress the signal intensity until they are depleted in the oxidation process. Pre-oxidized polymers give an appreciable light emission at elevated temperature in nitrogen, which may come from the fast decomposing peroxidic structures.

In the classical Boland – Gee scheme of hydrocarbon oxidation involving initiation,

1/ $PH \rightarrow P^{\cdot}$	w_i
2/ $P^{\cdot} + O_2 \rightarrow PO_2^{\cdot}$	k_2
3/ $PO_2^{\cdot} + PH \rightarrow POOH + P^{\cdot}$	k_3
4/ $POOH \rightarrow PO^{\cdot} + {}^{\cdot}OH$	k_{mono}
5/ $2\,POOH \rightarrow PO_2^{\cdot} + PO^{\cdot} + H_2O$	k_{bi}
6/ $PO^{\cdot} \rightarrow ketone + P^{\cdot}$	k_4
7/ $PO^{\cdot}\,({}^{\cdot}OH) + PH \rightarrow POH\,(H_2O) + P^{\cdot}$	k_5
8/ $2\,PO_2^{\cdot} \rightarrow POH + ketone^{*} + O_2^{*}$	k_6
9/ $PO_2^{\cdot} + P^{\cdot} \rightarrow POOP$	k_7
10/ $2\,P^{\cdot} \rightarrow P-P$	k_8
11/ $PO_2^{\cdot} + SH \rightarrow products$	k_{inh}
12/ $POOH + D \rightarrow products$	k_d

propagation and termination of free radical chain process, CL is assumed to arise in step 8), which schematically denotes the disproportionation of secondary or primary peroxyl radicals. In the case of hydroperoxide decomposition in a micro-domain, where its concentration is significantly above the average value, we also suspect the step 5), which is bimolecular decomposition of hydroperoxides. Formally it may be identified with explosion-like process potentially yielding excited particles as well. Ketone* and O_2^* are excited triplet states of ketone group and singlet oxygen, respectively. They are converted to the ground state either in a luminous way:

$$ketone^* \rightarrow ketone + h\nu$$
$$O_2^* \rightarrow O_2 + h\nu$$

or as heat dissipated in collisions.

One should be aware of the fact that only a very small part of excited states is converted to the ground state with the emission of the light, the prevailing part is quenched in a non-luminous way. This is a pre-requisite of more or less complex relation of observed CL intensity with the rate of initiation. Deconvolution of the respective initiation mechanism from CL - time runs may be rather difficult and sometimes impossible without a use of complementary methods. From the scheme of non-inhibited oxidation, assuming that chemiluminescence arises only in disproportionation of secondary or primary peroxyl radicals (step 8), one can propose that:

$$w_i + k_{mono}[POOH] + k_{bi}[POOH]^2 = k_6 [PO_2]^2 \approx I_{CL},$$

where I_{CL} is the intensity of chemiluminescence. If the initiation route is governed predominantly by the decomposition of hydroperoxides, i.e. $k_{mono}[POOH] + k_{bi}[POOH]^2 >> w_i$, the isothermal autocatalytic pattern of CL will be observed during polymer oxidation, while the CL intensity will decay steadily from some intial value for $k_{mono}[POOH] + k_{bi}[POOH]^2 << w_i$.

In the present paper we compare CL intensity – time runs for the oxidation of polypropylene, polyethylene and polyamide 6 (or polyamide 66) which are typical representatives of an autocatalytic increase of chemiluminescence intensity in time. However, the latter two polymers show more complex behavior, giving two mutually superimposed autocatalytic waves of chemiluminescence intensity runs.

Experimental

The chemiluminescence device Lumipol 2 was used for chemiluminescence measurements. The instrument was designed and manufactured at the Polymer Institute, Slovak Academy of Sciences, Bratislava, Slovak Republic. Lumipol 2 has the level of discrimination 2 counts/s at 40 °C. A polymer film (2-3 mg) having a diameter of 6 mm, or fine polymer powder, was placed on the aluminium pan of the diameter 9 mm to the oven of chemiluminescence device. Oxygen gas (flow 3.4 l/h) passed through the reactor during the measurement.

Polypropylenes were synthesized at the Institute of Science and Technology of Polymers, Madrid. Unstabilised polyamide 6,6 (Radilon A), containing only a small amount of phosphate (several ppm) as synthesis catalyst, and less than 0.5% weight residual oligomers, was kindly supplied by Radicinovacips SpA (Chignolo d'Isola, Italy). Polyethylene was Marlex 55180, free of stabilisers.

Results and Discussion

Kinetic analysis chemiluminescence – time runs

In our previous papers [3-6] we showed that the isothermal CL runs can be well fitted by the eq. 1)

$$I = \frac{A\exp(-k_1 t)}{[1 + Y\exp(-k_2 t)]^2} \qquad 1)$$

where Y is positive for the case of oxidation having the sigmoidal shape and negative for the case of oxidation showing the monotonous decay of chemiluminescence. The advantage of the equation above lies in its direct relation to the oxidation process.

It was formulated assuming the bimolecular reaction of polymer hydroperoxides as the initiation step of the reaction scheme extracted from the more general scheme in the Introduction:

$$2POOH \xrightarrow{k_{bi}} PO_2^{\cdot} + PO^{\cdot} + H_2O$$

$$PO^{\cdot} + PH \xrightarrow{k} POH + P^{\cdot}$$

$$P^{\cdot} + O_2 \xrightarrow{k_2} PO_2^{\cdot}$$

$$PO_2^{\cdot} + PH \xrightarrow{k_3} POOH + P^{\cdot}$$

$$2PO_2^{\cdot} \xrightarrow{k_6} products$$

Here we assume that the transfer reaction of alkoxyl radicals PO to polymer PH is considerably faster than the corresponding reaction of peroxyl radicals so that the alkoxyl radicals concentration can be neglected compared to that of peroxyl radicals. The Differential equations for concentrations of alkyl radicals and peroxyl radicals were converted to algebraic ones using the Bodenstein principle of steady state. For the time changes of hydroperoxide concentration [POOH] we ultimately obtain eq. 2.

$$[POOH] = \frac{X}{1 + Y \exp(-k_2 t)} \tag{2}$$

where $X = [POOH]_\infty$, $Y = \dfrac{[POOH]_\infty - [POOH]_0}{[POOH]_0}$

$[POOH]_\infty = \dfrac{k_4[PH]}{2\sqrt{k_6 k_{bi}}}$ and $k_2 = k_4[PH]\sqrt{\dfrac{k_{bi}}{k_6}}$

Having in mind the final goal to find an optimum fit with experimental runs we have postulated the relation between the CL intensity and time function F(t) so that it is the product of two terms. The first term represents the rate of hydroperoxide concentration changes. The second term, S(t), is the correction for the decay of the CL intensity when crossing the maximum, which is not so fast in experiments as it should correspond to the constant k_2 (See eq.4).

$$I = \mu F(t) = \mu (\frac{d[POOH]}{dt}) S(t) \tag{3}$$

(Proportionality constant μ is the yield of chemiluminescence.)

According to eq. 2 $\dfrac{d[POOH]}{dt} = \dfrac{k_2 XY \exp(-k_2 t)}{[1 + Y \exp(-k_2 t)]^2}$

$\hspace{11cm}$ 4)

when compared to the oxidation of uniform (liquid) medium from which we monitor the intensity of the light emission from a well defined surface, where all concentrations are well equilibrated, in solid polymers the situation is usually more complex. The diffusion of macromolecules is restricted and limits itself to polymer segments only. It may happen that the oxidation will start incidentally at some surface sites of the polymer from which it then spreads to the whole sample [7]. The surface emitting the light thus increases and so does the light of emission. The correction function S(t) takes such a possibility into account. In the first approximation we have expressed the function S(t) by an exponential rise as follows.

$$S(t) = S_0 \exp(rt)$$

$\hspace{11cm}$ 5)

where S_0 is the initial surface where the start of oxidation is potentially the most probable because of accumulation of different defect sites and r is the rate constant of oxidized surface spreading.

By combination of eqs. 3, 4 and 5 we finally obtain the analytical expression 1), in which $A = \mu k_2 XY S_0$ and $k_1 = k_2 - r$.

Some theoretical runs of CL intensity for a set of parameters: A=20000, $k_1 = 5 \ 10^{-6}$ s^{-1}, $k_2 = 1 \ 10^{-4}$ and several Y parameters are shown in Fig. 1. It can be seen that the induction time for the advanced stage of oxidation is determined not only by the corresponding rate constants but also by parameter Y, which is the ratio of maximum (at infinity time of the process) $[POOH]_\infty$ and initial concentrations of hydroperoxides $[POOH]_0$ reduced by 1. While the maximum concentration of hydroperoxides is determined by kinetic parameters and at a given temperature it should be constant, the initial concentration of hydroperoxides should be considered as the scale of the polymer quality including all other defect structures in the polymer due to processing, reactions of catalyst residuals, etc., which are converted to hydroperoxides. The higher the concentration of these defect structures is the lower the value of Y is. Provided that the defect structures predominate the maximum concentration of hydroperoxides kinetically attainable (Y<0), we face the case of steady decay of chemiluminescence intensity (See the decay curve in Fig. 1 for

A=1000, k_1=5 10^{-6} s^{-1}, k_2=1 10^{-4} s^{-1} and Y= -0.8). This may happen when, for example, the rate constant k_4 of the transfer reaction of peroxyl radicals to a polymer chain is very low and thus the lengths of the kinetic chains are rather short.

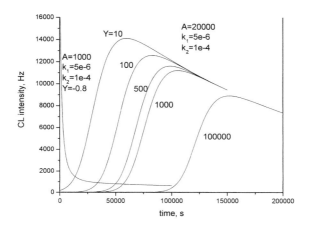

Figure 1.   Theoretical runs of chemiluminescence intensity derived from eq. 1 for different values of parameters Y

Experimental chemiluminescence intensity – time runs for non-stabilized and stabilized polypropylene

The experimental runs of chemiluminescence intensity in time for polypropylenes, prepared by Ziegler Natta and metallocene catalysts, having different properties including molar mass, the degree of isotacticity, etc., are shown in Fig. 2. The temperature of the experiment was 100 °C. From Fig. 2, the dependence on molar mass becomes quite obvious. The higher the molar mass, the higher is the induction time and lower is the maximum chemiluminescence intensity.

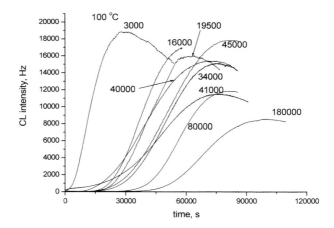

Figure 2. Experimental runs of chemiluminescence intensity – time for oxidation of powders of polypropylene (oxygen, 100 °C) – See also Table 1

The perfect fit of eq. 1 to experimental runs of CL from oxidized polypropylene was already demonstrated in our earlier papers [4,5]. Here we present parameters of the eq. 1 derived from such a fit (Table 1), corresponding to the series of experiments represented in Fig. 2.

It is of interest that parameter Y which is the scale of the concentration of defect structures in the polymer correlates well with the extent of isotacticity of the polymer (Fig. 3). The lowest concentration of the defect structures is in the polymer containing the highest isotacticity. 100% of isotactic structures from the mixture of isotactic, syndiotactic and atactic structures renders to polypropylene also the highest stability as seen from the induction times of oxidation (Fig. 4) and from rate constants of the fast decomposing peroxides (Fig. 5). Induction times were determined from the intercept of the straight line of maximum slope crossing the inflexion point of the CL run and time axis.

The conception of defect structures introduced through the term Y of eq. 1 may also be applicable to polypropylene stabilized by antioxidants. In Fig. 6 we see an example of the addition of

Table 1. Parameters of respective experimental runs and the fit by eq. 1 for oxidation of different polypropylenes (metallocene and Ziegler Natta) (powder, 3 mg) in oxygen at 100°C

Molar mass	% of isotacticity	I_{max}	t_{ind}	$(d[\frac{I}{I_{max}}]/dt)_{max}$ $* 10^5$	Y	$k_1 * 10^6$	k_2 x $*10^5$
Daltons		Hz	s	s^{-1}		s^{-1}	s^{-1}
3000	84.9	18806	4146	6.2	6.2	9.2	19
16000	86	17087	23639	3.6	46.1	-	12
19500	86.6	16027	25165	3.23	49.2	20	10
34000	92.5	15113	30990	2.92	47.3	10	8
40000 *	76.6	15417	19952	2.63	15.0	5.4	8
41000	89.4	11456	23661	2.48	16.3	20	6
45000 *	82.2	17952	29046	2.46	28.5	4.4	8
80000 *	100	11871	41290	3.61	274.1	6.4	11
180000 *	100	8568	50476	2.78	240.9	5.2	9

*Ziegler Natta polypropylenes

x rate constant of fast decomposing peroxides in polypropylene at 100°C is $8.7 \ 10^{-5} \ s^{-1}$, rate constant of slow decomposing peroxides in polypropylene at 100 °C is $2.4 \ 10^{-6} \ s^{-1}$ [9,10]

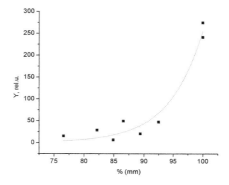

Figure 3.   The plot of the parameter Y from eq. 1 on the extent of isotacticity as determined by FTIR measurements

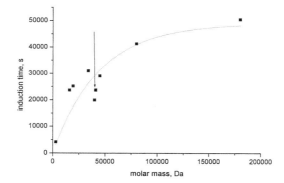

Figure 4.   The plot of the induction times from Fig. 2 versus molar mass of metallocene and (left from the arrow) Ziegler-Natta polypropylenes

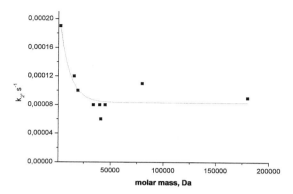

Figure 5.   The plot of the rate constant k_2 of eq. 1 determined from experimental runs of oxidation of polypropylene from Fig. 2

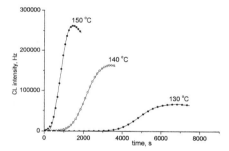

Figure 6.   Chemiluminescence runs for oxidation (oxygen) of polypropylene stabilized with 0.1 wt.-% of Irganox 1076. Lines are experimental records, points denote theoretical runs corresponding to the fit of experiment by eq. 1

0.1 % wt of hindered phenol Irganox 1076 to polypropylene supplied from another source (Chemie Linz, Austria). The presence of antioxidant increases the value of Y significantly indicating again that Y is the main factor determining the induction time of oxidation. It appears as though the concentration of defect structures in the polymer contributing to the initiation of oxidation is considerably reduced by the presence of antioxidant. It is of interest that the presence of Irganox 1076 also brings about the reduction of the rate constants k_2 and k_1 (Table 2). It is to be borne in mind that the constant k_2 coincides numerically with the rate constant of hydroperoxide decomposition in the advanced stage of oxidation.

Table 2.   Parameters of eq. 1 for chemiluminescence runs of oxidation of polypropylene stabilized with 0.1 wt. % of Irganox 1076

Temperature	A	Y	$10^5 * k_1$,	$10^3 * k2$,
°C	Hz		s^{-1}	s^{-1}
130	132 826	840.4	9	1.5
140	286 546	54.98	14	2.15
140*	62 843	5.51	17	3.9
150	465 570	17.69	34	4.13

*pure polypropylene

Comparison of chemiluminescence runs during oxidation of polypropylene, polyethylene and polyamides

On plotting chemiluminescence courses for the oxidation of the above three polymers at 140 °C we see that pure polypropylene attains the advanced stage of oxidation first, then followed by PA 66 and finally low density polyethylene (LD PE) (Fig. 7). On experimental runs of CL intensity – time of the two latter polymers, however, two well distinguished waves (I and II) are noticed. The initial intensity of CL accompanying the oxidation of polyamide 66 is, moreover, considerably higher than those of polypropylene (PP) and polyethylene. In one of our last papers [8] we have expressed an idea that this initial level of CL intensity corresponds to a faster oxidation of the terminal amino groups in polyamide 66. Regardless of the existence of two waves for the description of the kinetics of the process we have again used the eq.1 (Table 3). We see that the lowest concentration of defect structures (the highest Y) is found in polyethylene (Y=13.7), then in polypropylene (Y=5.31), while the highest concentration of defect structures exists in polyamide 66 (Y=1.83). This is probably due to the existence of terminal amino groups in the latter, which are potential sites of initial oxidizability. On the other hand, the rate constants k_2 are the highest for polypropylene, where oxidation occurs via tertiary hydroperoxides. The k_2 for polyamide 66 is one order of magnitude lower (oxidation takes place presumably via secondary hydroperoxides in the vicinity of the nitrogen of the amino group) and the lowest k_2 value was for the oxidation of polyethylene (oxidation occurs via secondary hydroperoxides).

Two-stage oxidation of polyethylene apparently disappears at temperatures below 130 °C (Fig. 8) and eq. 1 then describes the oxidation process very well. Parameters of the fit of CL from polyethylene at different temperatures are in Table 4. According to the definition of a respective term the higher values Y at lower temperatures correspond with higher maximum concentrations of hydroperoxides. The error brought about by application of eq. 1 to two waves reflects itself in a decrease of the rate constants k_2 and k_1 between 130 and 140 °C from the expected sequence.

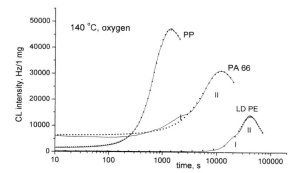

Figure 7.   Comparison of chemiluminescence intensity – time runs accompanying the oxidation of polypropylene, polyamide 6 and low-density polyethylene at 140 °C. Lines are experimental runs, points are theoretical fits by eq. 1

The effect of some salts (200 ppm) on polyamide 66 is shown in Fig. 9. The fit of experimental runs by eq. 1 again yields quite a good fit. However, the Y values convert to negative values when $CuCl_2$ is added. $CuCl_2$ performs a strong antioxidation action (Table 5).

Table 3. Parameters of eq. 1 for chemiluminescence intensity – time runs accompanying oxidation (oxygen) of polypropylene, polyethylene and polyamide 6 at 140 °C

Polymer	A	Y	$10^5 * k_1$	$10^4 * k_2$
	Hz		s^{-1}	s^{-1}
polypropylene	62 843	5.51	17	39
polyamide 6	50 945	1.83	3	2.8
polyethylene	66 686	13.57	3	1.1

Table 4. Parameters of eq. 1 for chemiluminescence kinetics for oxidation of polyethylene

Temperature	A	Y	$10^6 * k_1$	$10^6 * k_2$
°C	Hz		s^{-1}	s^{-1}
110	354 127	150.5	-	7.8
120	71 725	60	9.5	30
130	39 563	34	10	50
140	66 686	13.57	30	110
150	39 077	8.4	20	280

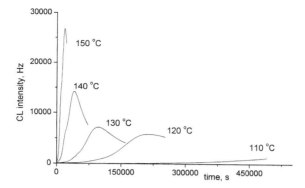

Figure 8.   Chemiluminescence intensity runs accompanying the oxidation (oxygen) of low-density polyethylene

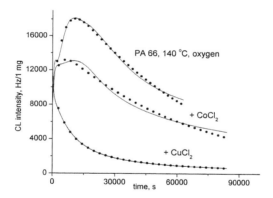

Figure 9.   Chemiluminescence intensity – time runs for oxidation of polyamide 66 (oxygen) at 140°C in the presence of 200 ppm of some salts.

Table 5.   The parameters of eq. 1 for chemiluminescence runs accompanying the oxidation of polyamide 66 (PA 66) at 140 °C

System	A	Y	$10^5 * k_1$	$10^4 * k_2$
	Hz		s^{-1}	s^{-1}
PA 66 pure	23 263	0.535	2	2.5
PA 66 + CoCl$_2$	15 097	0.227	2	3.8
PA 66 + CuCl$_2$	1 397	-0.631	1	0.3

The rate constants k_2 of PA66 + CuCl$_2$ are almost one order of magnitude lower than those for pure polyamide 66. Accordingly, CoCl$_2$ acts as a pro-oxidant.

Conclusions

The kinetics of chemiluminescence intensity – time runs may well be fitted by eq. 1 in the case of polypropylene and polyethylene provided that both polymers give a single autoaccelarating increase of the light emission. Oxidation of polyethylene above 130 °C and polyamides is accompanied typically by two waves chemiluminescence kinetics. From the kinetic equation the concentration of defect structures initially present in a polymer are related to the initial concentration of hydroperoxides. Kinetic parameters found from eq. 1 are well related to the rate constants of hydroperoxides decomposition during polymer oxidation.

Acknowledgments

We are grateful to Grant agency VEGA of Slovak Republic for the financial support of project No. 2/1127/21 due to which this paper was realized. We acknowledge also the support from Comisión Interministeral de Ciencia y Tecnología. Our thanks are due to P. Cerruti from University, Napoli, Italy for the supply of polyamide 66 samples.

[1]   G. E. Ashby, *J. Polym. Sci.* **1961**, *50*, 99.
[2]   [2a] M. P. Schard, C. A. Russell, *J. Appl. Polym. Sci.* **1964**, *8*, 985;[2b] M. P. Schard, C. A. Russell, *J. Appl. Polym. Sci.* **1964**, *8*, 997.
[3]   L. Matisová-Rychlá, J. Rychlý, *Polym. Degrad. Stab.* **2000**, *67*, 515.
[4]   J. Rychlý, L. Matisová-Rychlá, D. Jurčák, *Polymer Degrad. Stab.* **2000**, *68*, 239.
[5]   L. Rychlá, J. Rychlý, "New concepts in chemiluminescence for the evaluation of thermooxidative stability of polypropylene from isothermal and non-isothermal experiments", in: *Polymer Analysis and Degradation*, A. Jimenez, G. E. Zaikov, Eds., Nova Science Publishers, New York 2000, p.124.
[6]   J. Rychlý, L. Matisová-Rychlá, P. Tiemblo, J. Gomez-Elvira, *Polym. Degrad. Stab.*, **2001**, *71*, 253.
[7]   G. George, M. Celina, "Homogeneous and heterogeneous oxidation of polypropylene", in: *Handbook of Polymer Degradation*, 2nd edition, S. Halim Hamid, Ed., Marcel Dekker, New York 2000, p. 277.
[8]   P. Cerruti, C. Carfagna, J. Rychlý, L.Matisová-Rychlá, *Polym. Degrad. Stab.*, submitted.
[9]   J. C. W. Chien, "Hydroperoxides in degradation and stabilization of polymers", in: *Polymer Stabilisation*, Wiley, New York 1972, Chapter 5, p. 95.
[10]   J. C. W. Chien, H. Jabloner, *J.Polym. Sci. A-1* **1968**, *6*, 393.

Macromol. Symp. **2004**, *214*, 279-288

Influence of Reactive Compatibilization on the Morphology of Polypropylene/Polystyrene Blends

J. Pionteck,* P. Pötschke, U. Schulze, N. Proske, A. Kaya, H. Zhao, H. Malz

Institute of Polymer Research Dresden, Hohe Straße 6, D-01069 Dresden, Germany

E-mail: pionteck@ipfdd.de

Summary: To study the efficiency of different mechanisms for reactive compatibilization of polypropylene/polystyrene blends (PP/PS blends), main chain or terminal-functionalized PP and terminal-functionalized PS have been synthesized by different methods. While the *in-situ* block and graft copolymer formation results in finer phase morphologies compared to the corresponding non-reactive blends, the morphology development in the ternary blend system PP/PS + HBP (hyperbranched polymer) is a very complex process. HBP with carboxylic acid end groups reacts preferably with the reactive sites of the oxazoline functionalized PS (PS-Ox) and locates mainly within the dispersed PS-Ox phase. A bimodal size distribution of the PS-Ox particles within the oxazoline modified PP (PP-Ox) matrix phase is observed with big PS-Ox particles (containing the HBP as dispersed phase) and small PS-Ox particles similar in size to the unimodal distributed particles in the non-reactive PP-Ox/PS-Ox blends. Factors influencing the morphology are discussed.

Keywords: blends; morphology

Introduction

The compatibilization of immiscible blends can be done by the addition of emulsifiers (so called compatibilizers) as third component or by the use of reactive polymers, which can create block or graft copolymers at the interface during the melt mixing process. The *in-situ* formed compatibilizers reduce the interfacial tension, enhance the interfacial adhesion, stabilize the interphase during melt processing, and reduce in this way the tendency of coalescence, resulting in finer dispersions compared to non-reactive blends. This can lead to an improvement in the mechanical and other properties of multiphase blends. While the mechanisms of block- and graft copolymer formation have been widely studied over the past several years, a third mechanism, the *in-situ* formation of mixed star copolymers at the interphase of two immiscible polymers, due to the addition of highly functional hyperbranched polyesters (HBP), is a rather new approach to compatibilization. Hannerfeldt et al. [1-3] investigated the grafting of aliphatic HBP onto a polypropylene (PP) chain. In

DOI: 10.1002/masy.200451020

this way the number of reactive sites per molecule is increased. The effect of this grafted PP blends with polyamide 6 on interfacial properties and blend morphology was reported.

However, the morphology of the reactive blends depends not only on the reactivity of the ble[?] system, but also on physical properties such as the viscosities and the interfacial energies of t[?] blend components, as has been shown by Pötschke et al. [4,5]

To compare the different reactive compatibilization mechanisms (Scheme 1) and to separate t[?] physical factors from the effect of copolymer formation we have compared morphologies of no[?] reactive and reactive blends (which are able to form graft, block, or star-like copolymers duri[?] melt mixing), considering the viscosity ratios and changes in interfacial tensions due to t[?] chemical modification. For these studies we have chosen the blend syste[?] polypropylene/polystyrene (PP/PS), which may be considered as a model for PS/elastomer blen[?] Advantageous for PP and PS is that there exist different chemical pathways by which to obta[?] main chain or endgroup functionalised polymers. As model reaction for the compatibiliz[?] formation the conversion of 1,3-oxazolines with carboxylic acid, which is known as fast a[?] selective, has been used.

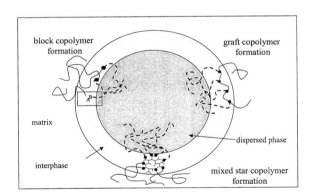

Scheme 1. Formation of block, graft, or star copolymers at the interface between immisci[?] polymers during melt mixing; A, B: different functionalities; black spots: coupl[?] sites [9]

Materials – Synthetic Strategies

The necessary end group and main chain functionalized reactive PP and PS were synthesised [?] different synthetic pathways. Their characterisation is given in the corresponding citations, if [?] mentioned here.

he PS were synthesized by the TEMPO method. The initiator used for the preparation of the arboxylic acid terminated PS (**PS-COOH**) was the commercially available 4,4'-azo-bis-(4-yanopentanoic acid) (ABCPA) [6-8]. The non-reactive PS was prepared using 2,2'-azo-bis-obutyric acid nitrile (AIBN). To stabilize the product the TEMPO group was removed by xidation with m-chloroperbenzoic acid [7]. Oxazoline-terminated PS (**PS-Ox**) was obtained nalogously but the ABCPA initiator was converted into an bisoxazoline by conversion the BCPA-chloride with (4-hydroxyphenyl)-1,3-oxazoline [7,9]. The degree of the PS unctionalization was about 75 %. An alternative route to obtain PS-Ox is the atom transfer radical olymerization (ATRP) [7,10,11] which results in high functionalities, but the remaining bromide end roup may result in unfavourable side reactions.

P grafted with oxazoline (**PP-g-Ox**) was prepared by grafting of PP (Novolen 1127 MX, BASF, ermany) with ricinol oxazoline maleinate (Henkel, Germany). The oxazoline content of the PP-g-x was 1.3 wt% [8]. Alternatively, main chain oxazoline-functionalised PP was obtained by opolymerisation of propene with different unsaturated 1,3-oxazolines [12,13]. The detailed structure nd analysis of the **PP-co-Ox** used in the blends is given in [13].

arboxylic terminated PP (**PP-COOH**) with different molecular weights and therefore viscosities ould be obtained by hydrosilylation of peroxidic degraded PP with 4-pentenoic acid [7]. The entenoic acid had to be used in the trimethylsilyl ester form to suppress the poisoning effect of the ee acid on the platinum catalyst. The peroxidic degraded PP (and therefore also the hydrosilylation roduct PP-SOOH) is a mixture of monofunctional, bifunctional and nonfunctional polymer chains. he hydrosilylation with oxazoline monomers is not possible due to side reactions of the ydrosilylation agent with the oxazoline ring.

o achieve better control over the functionality, metallocene catalysed PP, which obtains exactly ne unsaturated endgroup per polymer chain, may be used. Such PP with an molecular weight of $M_n = 21000$ g/mol with a polydispersity (PD) of 2.05, was used to prepare oxazoline-terminated PP **PP-Ox**) by a hydroboration reaction. After the hydroboration with 9-borabicyclo[3.3.1]nonane, naleic acid anhydride can be coupled to the PP in the presence of oxygen. The resulting maleic rminated PP offers the opportunity for a large variety of further modification reactions [14]. We repared PP-Ox by its conversion with 2-(4-aminophenyl)-1,3-oxazoline under the same conditions s described by [14]. During the reactions no changes in the molecular weight could be observed. lowever, the conversion was incomplete. We calculated a functionality of 35 %, that means about very 3rd PP chain contained one 1,3-oxazoline endgroup.

Polyfunctional hyperbranched polyester (HBP-COOH, Scheme 2) with Mn = 26000 g/mol wa obtained by polycondensation of the AB$_2$ type monomer 1,5-bis-(4´-carboxyphenyl)pentan-3-c Details are given in [9].

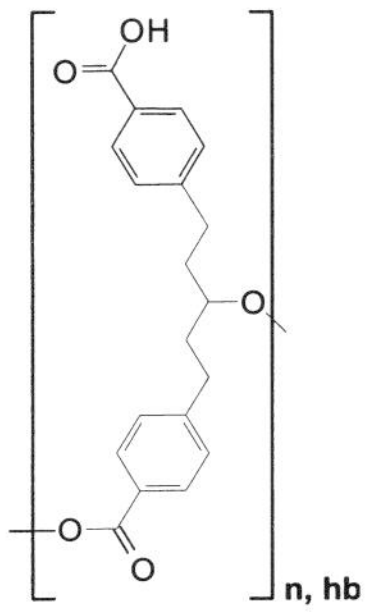

Scheme 2. Chemical Structure of HBP-COOH

Methods

All blends were prepared by melt mixing at 200 °C. Most of the blends forming graft- or bloc: copolymers. A Miniatur Mixing Reactor EK-3-5C (NCS, Japan) was used (rotating piston in fixed cylinder with 4 shear edges in the bottom; 3 g; shear rate about 6-10 s^{-1}; 30 min. – if n mentioned otherwise). The blends in the composition PP/PS = 2/1 (by weight) plus 5 wt-% HE and the blends based on PP-co-Ox were prepared by melt mixing using a Micro Compound (DACA Instruments; two conical co-rotating screws with a bypass allowing the material to circula for defined periods; capacity of 4.5 cm^3) at 200 °C with 100 rpm for 5 min, if not otherwi mentioned. The shear rate was much higher than in the other mixing equipment (similar to she rates in conventional twin screw extruders) but could not be quantified.

The particle size of the PS was investigated by SEM of cut surfaces after chemically etching the F with xylene or THF for 3 h. A SEM LEO 435 VP (Leo Elektronenmikroskopie, Germany), with a acceleration voltage of 10 kV, was used. Selective etching of the HBP phase was performed wi NaOH/water at room temperature for 4 h.

Melt rheology was investigated using a ARES rheometer (Rheometrics, USA) in a nitroge atmosphere. Frequency sweeps were performed in the linear viscoelastic range at 200°C using parallel plate geometry. Surface and interfacial tensions of polymer melts were measured using tl pendant drop method, using a self made apparatus. SEC measurements of PS samples we

erformed at ambient temperature using modular chromatographic equipment with a refractive
dex detector. The molecular weights of PP samples were determined using a PL-GPC 210
Polymer Laboratories, UK) at 135°C with 1,2,4-trichlorbenzene (TCB).

esults and Discussion

he morphology development when melt mixing immiscible polymers is a combined process of
op deformation and break up on the one hand and coalescence on the other hand. From the
aylor-Equation given below, the minimum particle diameter d_T which can be reached by mixing of
vo immiscible fluids in dependent on the shear rate $\dot{\gamma}$, matrix viscosity η_m, viscosity ratio between
spersed and matrix phase λ, and interfacial tension σ_{12} can be calculated. Even if this relation is
alid only for Newtonian fluids, this relation can be used for the estimation of the smallest
chievable diameter in polymer mixtures. But the same factors influencing the drop deformation
id break up also influence the coalescence, which coarsens the morphology. However, there exists
o good theoretical model for the description of all the correlations during the process of
palescence. In addition, with increasing dispersed phase content the coalescence probability
ncreases strongly. Therefore only morphologies with similar compositions can be compared with
ach other.

$$d_T = \frac{4 \cdot \sigma_{12} \cdot (\lambda + 1)}{\dot{\gamma} \cdot \eta_m \cdot \left[\left(\frac{19}{4} \cdot \lambda \right) + 4 \right]}$$ Taylor-Equation

[8] we showed that the interfacial and surface tensions of polymers are not, or only slightly,
ffected by reactive modification. Therefore, when comparing reactive with non-reactive blend
/stems, changes in the morphology can be caused only due to the chemical reactions. When no
eactions occur then the interfacial tension will be constant. Otherwise we noticed a strong effect of
e reactive sites on the viscosity of the polymers, even if the molecular weight is not changed
uring the process of chemical modification [9]. Therefore, the viscosity ratio has to be determined
or each polymer combination.

he particle size is smaller in all in-situ compatibilized blends. It was observed that with increasing
S content in non-reactive blends a significant increase in the particle size and particle size
istribution, caused by coalescence, occurs, which is less pronounced in the reactive blends. Fig. 1
ives an example of the influence of reactivity on the final morphology when forming block-

copolymers in-situ [7]. At unfavourable viscosity ratios, λ greater than 1, the effect of the reactivity clearly visible, while at favourable viscosity ratios, λ near 1, the reactive system has only a slight improved phase morphology compared to the non-reactive system. The favourable viscosity rat results in even finer PS distribution than in the reactive system with more different viscosities. this case the physical factors dominate the blend morphology.

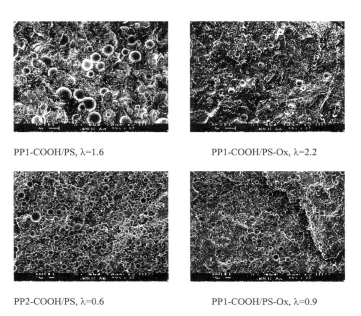

PP1-COOH/PS, λ=1.6

PP1-COOH/PS-Ox, λ=2.2

PP2-COOH/PS, λ=0.6

PP1-COOH/PS-Ox, λ=0.9

Figure 1.   Influence of the block-copolymer formation and viscosity on the morphology of PP/PS = 70/30 blends, SEM of cryofractures (frame size 310x220 μm) [7]

The in-situ graft-copolymer formation seems to have a stronger influence on the morpholog probably due to the higher content of reactive sites in graft copolymers compared to endgrou modified polymers. The influence of a changed viscosity ratio is clearly visible when comparin blends prepared with PS-COOH 2 (Fig. 2) [8]. The mean number average particle diameter d_n f this reactive blend with a viscosity ratio of about 1 is much higher than in the reactive blend wi the lower viscosity ratio but still much finer than in the non-reactive blend with the low viscosi ratio.   It must be noted that the functionality is also changed;   PS-COOH 2 contains much few reactive sites than PS-COOH 1 does.

In order to study the influence of the degree of functionalization on the morphology, a series of P g-Ox/PS = 70/30 blends were prepared. PS-COOH1 was mixed in different compositions with t non-reactive PS. In this way the ratio between carboxylic acid groups and oxazoline groups w

aried between 0.05 and 0.51 whilst maintaining a constant viscosity ratio. Micrographs are shown
1 Figure 3. The results show that the addition of more than 50 wt% of modified PS led to no further
improvement in the dispersity. This means that complete functionalization of the PS is not
ecessary for the in-situ compatibilization. This is an important fact since polymers in which every
hain contains reactive sites are accessible only by a limited number of methods.

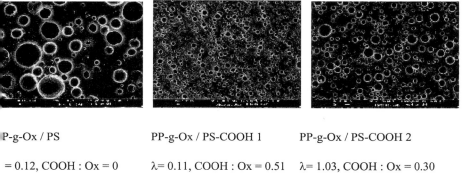

P-g-Ox / PS   PP-g-Ox / PS-COOH 1   PP-g-Ox / PS-COOH 2

= 0.12, COOH : Ox = 0   λ= 0.11, COOH : Ox = 0.51   λ= 1.03, COOH : Ox = 0.30

igure 2. Influence of the graft-copolymer formation and viscosity on the morphology of PP/PS =
          70/30 blends, SEM of cryofractures (frame size 64x48 μm)

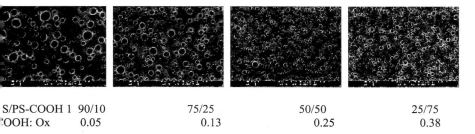

S/PS-COOH 1 90/10   75/25   50/50   25/75
'OOH: Ox   0.05   0.13   0.25   0.38

igure 3. SEM of etched surfaces of PP-g-Ox/PS = 70/30 blends with varying PS/PS-COOH 1
          ratio (frame size 64x48 μm)

'ontrary to this finding we found in another system based on PP-co-Ox/PS-COOH a continuous
hange in the morphology with an increasing amount of reactive polymers. When the nonreactive
P was completely replaced by the reactive copolymer even phase inversion was observed
Fig. 4). [13]

1ore complex is the situation when trying to use HBP-COOH as compatibilizing agent in blends
ased on PS-Ox and PP-Ox [9]. When melt mixing PP with PS and PP-Ox with PS-Ox (in both

blends the components cannot react with each other) the PP-Ox/PS-Ox blend has a much fine morphology that the PP/PS blend. The reason is the changed viscosity ratio, even if the molecula weights of the reactive polymers are nearly equal to those of the non-reactive polymers. Th addition of HBP-COOH which should form compatibilizing mixed stars at the interphase, due to th reaction with both oxazoline-modified polymers, did not result in a finer morphology. In contras very big holes were observed after etching with THF which is a solvent for the PS and HBP (Fig.5)

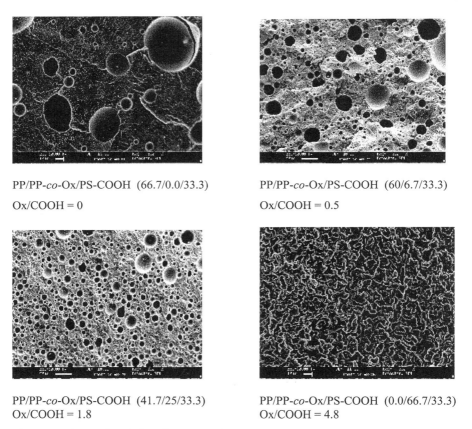

PP/PP-*co*-Ox/PS-COOH (66.7/0.0/33.3)
Ox/COOH = 0

PP/PP-*co*-Ox/PS-COOH (60/6.7/33.3)
Ox/COOH = 0.5

PP/PP-*co*-Ox/PS-COOH (41.7/25/33.3)
Ox/COOH = 1.8

PP/PP-*co*-Ox/PS-COOH (0.0/66.7/33.3)
Ox/COOH = 4.8

Figure 4.   SEM of etched surfaces of PP/PP-co-Ox/PS-COOH blends with varying PP/PP-co-O ratios (frame size 233x187 µm)

The reason for this a typical behaviour is the much better compatibility of the HBP to PS and PS-O than that to PP or PP-Ox. HBP is immiscible with both polymers but, due to the aromatic characte the HBP forms much smaller particles in PS than in PP. In reactive PP-Ox the HBP-COO distribution is somehow improved but still much more worse than in PS or PS-Ox. Therefore, whe mixing all components together the HBP locates favourably within the PS phase, as well in reacti\

s in non-reactive blends. This results in an increase in the PS-Ox viscosity, which finally causes
ιe formation of big PS-particles filled with rather finely distributed HBP. The size of the filled PS
ιrticles increases with time. Beside this PS phase another part of PS which contains no HBP
ιclusions has about the same size as it does in the blend free of HBP. Etching the three-component
lend with NaOH/water dissolves only the HBP phase and reveals the phase in phase morphology
ᵢig.6).

PP/PS (λ= 6.8)

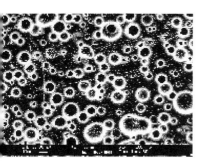 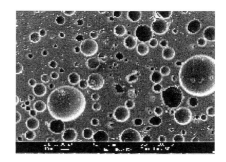

PP-Ox/PS-Ox (λ= 3.0)

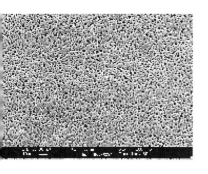 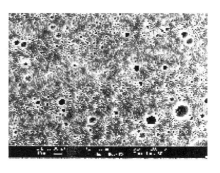

no HBP   5 wt% HBP-COOH

igure 5.   SEM of etched surfaces of PP/ /PS blends (2/1) with and without addition of 5 wt%
           HBP-COOH etched with THF (frame size 240x180 μm)

288

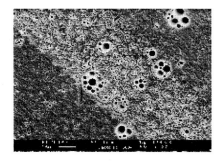 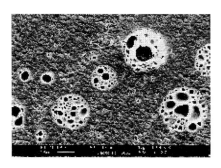

5 min. mixed   30 min. mixed

Figure 6.   Time dependence of the three-phase morphology of PP-Ox/ /PS-Ox (2/1) + 5 wt% HP
COOH etched with NaOH/water (frame size 116x86 μm)

The phase in phase morphology has similarities to the structure of high impact PS (HIP
Unfortunately, the rather low molecular weights of our polymers result in very brittle blends a
mechanical properties have not been determined.

Acknowledgement

The authors are grateful for the support of this work by the Deutsche Forschungsgemeinsch
(Collaborative Research Centre 287) and the Max-Buchner-Forschungsstiftung.

[1]   M. Baumert, R. Mülhaupt, *Macromol. Rapid. Commun.* **1997**, *18*, 787.
[2]   G. Jannerfeldt, L. Boogh, J.-A. E. Manson, *J. Polym. Sci.: Polym. Phys.* **1999**, *37*, 2069.
[3]   G. Jannerfeldt, L. Boogh, J.-A. E. Manson, *Polymer* **2000-1**, *41*, 7627.
[4]   G. Jannerfeldt, L. Boogh, J.-A. E. Manson, *Polym. Eng. Sci.* **2000-2**, *41*, 293.
[5]   A. Kaya, L. Jakisch, H. Komber, D. Voigt, J. Pionteck, B. Voit, U. Schulze, *Macromol. Rapid Comm.* **2001**, *22*, 972.
[6]   A. Kaya, G. Pompe, U. Schulze, B. Voit, J. Pionteck, *J. Appl. Polym. Sci.* **2002**, *86*, 2174.
[7]   H. Malz, H. Komber, D. Voigt, J. Pionteck, *Macromol. Chem. Phys.* **1998**, *199*, 583.
[8]   H. Malz, *"Synthesis of End-functionalized Polystyrenes and Polypropylenes"*, PhD Thesis, Technische Universit
Dresden, Der Andere Verlag, Bad Iburg 1999.
[9]   H. Malz, H. Komber, D. Voigt, I. Hopfe, J. Pionteck, *Macromol. Chem. Phys.* **1999**, *200*, 642.
[10] H. Malz, J. Pionteck, P. Pötschke, H. Komber, D. Voigt, J. Luston, F. Böhme, *Macromol. Chem. Phys.* **2001**, *202*, 2148.
[11] J. Pionteck, P. Pötschke, N. Proske, H. Zhao, H. Malz, D. Beyerlein, U. Schulze, B. Voit, *Macromol. Symp.* **2003** (in press).
[12] P. Pötschke, K. Wallheinke, H. Stutz, H. Fritsche, *J. Appl. Polym. Sci.* **1997**, *64*, 749.
[13] P. Pötschke, K. Wallheinke, H. Stutz, *Polym. Eng. Sci.* **1999**, *39*, 1035.
[14] P. Pötschke, H. Malz, J. Pionteck, *Macromol. Symp.* **2000**, *149*, 231.
[15] H. Zhao, J. Piontek, C. Taesler, P. Pötschke, *Macromol. Chem. Phys.* **2001**, *202*, 313.

Side Chain Extension of Polypropylene by Aliphatic Diamine and Isocyanate

K. Y. Kim, S. C. Kim*

Center for Advanced Functional Polymers, Korea Advanced Institute of Science and Technology, 373-1, Gusong-dong, Yusong-gu, Taejon 305-701, Korea
E-mail : kimsc@mail.kaist.ac.kr

Summary: A post-reaction modification of polypropylene by which to obtain increased melt strength is by reactive extrusion of randomly functionalized PP with polyfunctional monomers. In this study, reactive melt processing of maleated polypropylene (PP-g-MA) with poly[methylene (phenylene isocyanate)] (PMPI), hexamethylene diamine (HMDA), and poly(propylene glycol)-bis-(2-propylamine) (Jeffamine D-400) was carried out. The resulting chain-extended PP-g-MA was confirmed by FTIR analysis. Its thermal and rheological properties were also measured. HMDA and D-400 react with both the carboxylic acid and anhydride form of MA. Though PMPI only reacts with the carboxylic acid of MA, PP-g-MA reacted with PMPI has higher gel content and storage modulus compared to PP-g-MA reacted with the amine. This is because of the higher functionality of PMPI and localized reaction between the isocyanate in PMPI and the carboxylic acid in PP-g-MA.

Keywords: amine; chain extension; isocyanate; maleated polypropylene; melt strength

Introduction

Polypropylene (PP), one of the most widely used thermoplastics, possesses excellent properties, such as high melting temperature, high tensile modulus, low density and good chemical resistance. But PP has limitations in applications such as foaming, thermoforming and extrusion coating due to its poor melt strength. There have, therefore, been many attempts to overcome this disadvantage. The introduction of long-chain branching onto the PP backbone is one possible way in which the melt strength can be improved. Some types of post-reaction modifications have included increasing the melt strength of PP by introducing long-chain branching. Scheve et al.[1] tried to achieve this objective by irradiating a solid PP with high energy radiation. Many researchers, including Wang et al.[2], have shown that branching can be introduced by means of the recombination reaction in which PP reacts with polyfunctional monomers, having more than two double bonds, in the presence of peroxide. Another possibility is by the reaction between a randomly functionalized PP and polyfunctional monomers. In this study maleated polypropylene

(PP-g-MA) was used as a randomly functionalized PP, as were two types of polyfunctional monomer: polymeric isocyanate (PMPI) and diamine (HMDA, Jeffamine D-400).

Experimental

Materials

PP-g-MA with a MA content of 2 wt% (Honam Petrochemical Corporation) was used. The weight average molecular weight was 23,000 g/mol and the polydispersity index 2.3. Poly[methylene (phenylene isocyanate)] (PMPI), hexamethylene diamine (HMDA), and poly(propylene glycol)-bis-(2-propylamine) (trade name Jeffamine D-400) were used as chain extenders. The chemical structures of the chain extenders are shown in Figure 1.

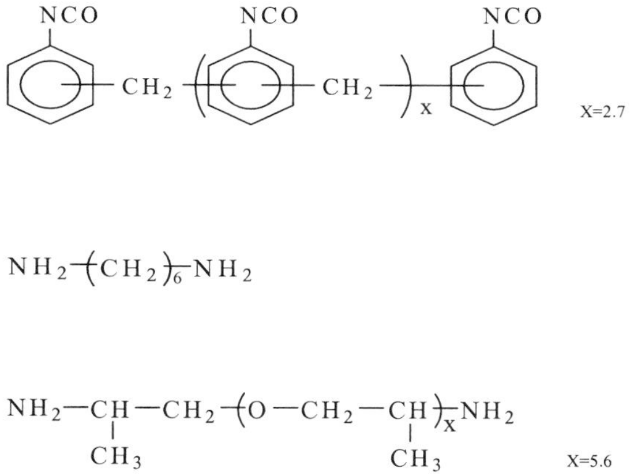

PMPI

HMDA

$$NH_2 \!-\!\!\left(CH_2\right)_{\!6}\!\!-\!NH_2$$

D-400

$$NH_2\!-\!\underset{CH_3}{\underset{|}{CH}}\!-\!CH_2\!-\!\!\left(O\!-\!CH_2\!-\!\underset{CH_3}{\underset{|}{CH}}\right)_{\!x}\!\!NH_2 \qquad X\!=\!5.6$$

Figure 1.   Chemical structures of the chain extenders used in this study.

Sample preparation and purification

PP-g-MA and chain extenders were reacted in a small Mini Max mixer at 175^0C for 10 minutes. The shear rate was 5 s^{-1}. One gram of PP-g-MA was charged into a preheated mixer and sheared. After the charged polymer had melted, the chain extender was added. Samples with different molar ratios of the reactive group for each chain extender were produced. The molar ratios of functional groups and weight fractions of chain the extenders are listed in Table 1. All the samples were purified by extracting the reaction product in the form of powder for 48 hours with an appropriate solvent. The polymers reacted with PMPI were purified by Soxhlet extraction with boiling acetone, and the polymers reacted with HMDA and D-400 with boiling methanol. After extraction, the polymers were washed with water and dried for 6 hours at 60^0C in a vacuum oven.

Table 1.   Molar ratios of functional groups and weight fractions of chain extenders.

	Sample name	Molar ratio of functional groups[a]	Weigt fraction[b] (wt %)
PMPI	PMPI 1/4	1/4	0.39
	PMPI 1/2	1/2	0.78
	PMPI 1/1	1/1	1.56
HMDA	HMDA 1/4	1/4	0.30
	HMDA 1/2	1/2	0.59
	HMDA 1/1	1/1	1.18
D-400	D-400 1/4	1/4	1.02
	D-400 1/2	1/2	2.04
	D-400 1/1	1/1	4.08

[a] [NCO]/[MA] for PMPI, and [NH$_2$]/[MA] for HMDA and D-400

[b] Weight ratio of chain extender to PP-g-MA

Measurements

FTIR analyses were done using a Jasco FT/IR-480 plus. Thin films for FTIR measurement were made by pressing at 200^0C. Thermal properties of 5-10 mg samples were studied by differential scanning calorimetry (DSC), using a Seiko Extar 6000 DSC. The temperature scanning rate was

10 ^{0}C/min for both heating and cooling steps. Gel content was measured according to the ASTM D2765-84 method. Rheological properties were measured using an ARES rheometer of Rheometric Scientific, Inc. A parallel plate was used, and the gap size was 1mm.

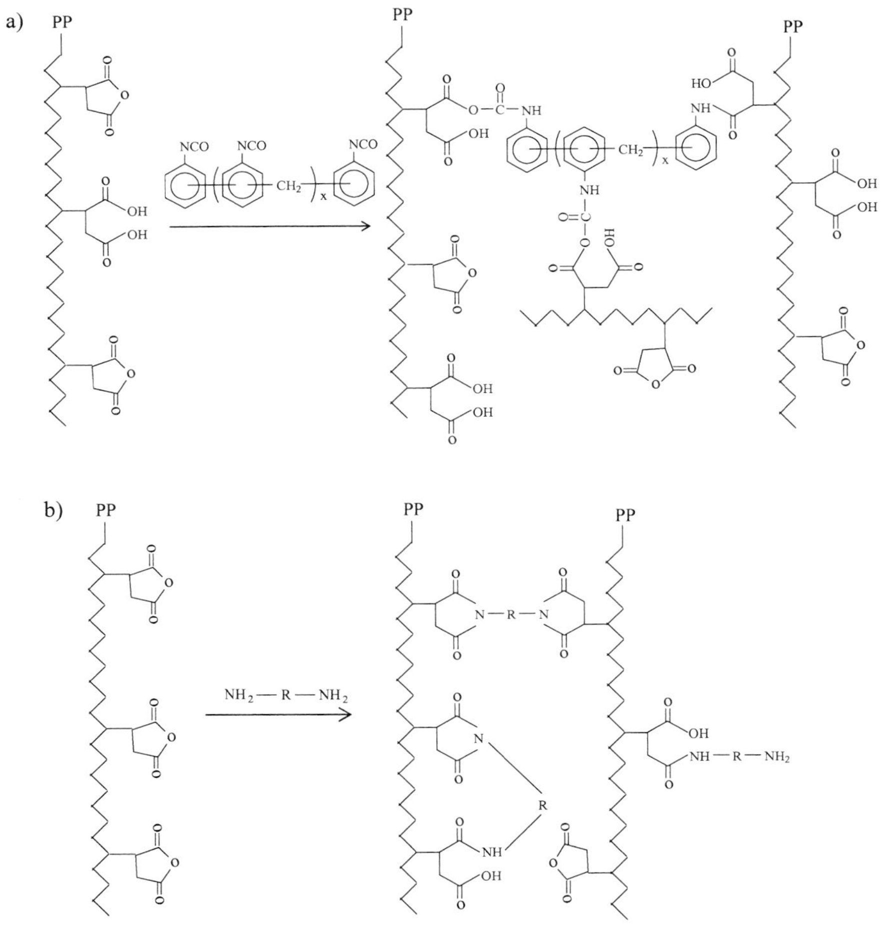

Figure 2. Simplified reaction mechanisms of reaction between a) PP-g-MA and PMPI, and b) PP-g-MA and primary diamine.

Results and Discussion

Chemistry

Some portions of MA in PP-g-MA exist in the form of anhydride and the others in the form of hydrolyzed free acid. As it is reported that an isocyanate group has good reactivity with carboxyl and hydroxyl groups, PP-g-MA can be chain-extended by reaction with PMPI[3]. The reaction mechanism of an isocyanate with a carboxyl group is shown in Figure 2a. In the case of the reaction between an isocyanate and a primary amine, the former can react with both the anhydride form of MA and the hydrolyzed free acid form of MA, to form an imide ring through the reaction mechanism shown Figure 2b.[4-5]

FTIR analysis

FT-IR spectra of chain-extended PP-g-MA and unreacted PP-g-MA are shown in Figure 3. The FT-IR spectra indicated that the MA in PP-g-MA exists in the forms of anhydride (1850 and 1780cm^{-1}) and hydrolyzed free acid (1712cm^{-1}). For samples reacted with PMPI, a decrease in intensity for the absorption peak at 1712cm^{-1} is observed, with the appearance of new peaks at 1595 and 1512cm^{-1}. Further, the intensity of these new peaks increases with increasing PMPI content. These new peaks indicate that an urea linkage was formed by the reaction between hydrolyzed free acid in the MA and isocyanate group in the PMPI. No change in the intensity of the peaks at 1850 and 1780 cm^{-1} means that isocyanate group did not react with the anhydride form of MA. For samples reacted with HMDA, characteristic peaks of MA at 1850, 1780cm^{-1} and 1712cm^{-1} disappear and new absorption peaks at 1770 and 1702cm^{-1}, that may be corresponding to the carbonyl group of the amic acid or imide, appeared. In the spectrum of samples reacted with D-400, the changes in characteristic absorption peaks of MA are similar to those of the HMDA series. Although characteristic peaks of MA still remain in the spectrum of D-400, absorption peaks at 1770 and 1702cm^{-1} increase with an increasing amount of D-400. Yet another new peak, at 1104cm^{-1}, attributed to the C-O-C stretching, increases with D-400 content.

a)

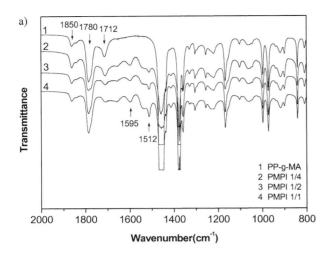

b)

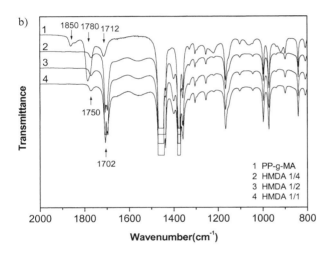

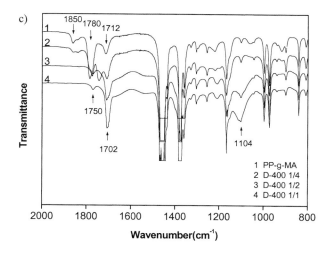

Figure 3. FTIR spectra of PP-g-MA and PP-g-MA reacted with a) PMPI, b) HMDA, and c) D-400.

Thermal properties

In Table 2, melting temperatures and ΔH_m of all samples measured by DSC are tabulated. Unreacted PP-g-MA has two melting points: 148.88 ^{0}C and 154.52 ^{0}C. Melting temperatures of all chain-extended samples are lower than the melting temperature of PP-g-MA, and reduced ΔH_m values are measured. Although PMPI only reacts with carboxylic acid, the decrease in ΔH_m is high even at the lowest molar ratio of PMPI. These decreases in ΔH_m for PP-g-MA reacted with PMPI are attributed to the high gel content, shown in Table 2. The average functionality of a PMPI molecule is 4.7 and is higher than those of amines used in this study. The reaction between PMPI and PP-g-MA can be localized, since the maleic anhydride group is hydrolyzed in local areas where moisture is present, and PMPI can only react with the carboxylic acid form of MA. Hence, PP-g-MA reacted with PMPI has higher gel content than PP-g-MA reacted with amine at the same molar ratio. For amine-extended samples, with increasing amine content, gel content increases and ΔH_m decreases.

Table 2.   Thermal properties and gel content.

	T_{m1}	T_{m2}	ΔH_m	Gel content
	°C	°C	J/g	wt%
PP-g-MA	148.88	154.52	94.45	0
PMPI 1/4		151.26	77.34	7.2
PMPI 1/2		150.96	77.97	21.4
PMPI 1/1		152.16	79.82	25.3
HMDA 1/4		149.98	82.46	2.1
HMDA 1/2	133.09	144.76	75.36	5.8
HMDA 1/1	132.93	143.55	71.10	22.1
D-400 1/4		150.20	92.78	1.2
D-400 1/2		145.84	81.04	2.6
D-400 1/1		146.26	77.71	15.7

Rheological properties

Chain-extended PP-g-MA with high gel content has very high normal stress under the conditions where rheological properties were measured, hence rheological properties could not be measured for samples with gel content greater than 10 wt%. Storage moduli of D-400 1/4 and HMDA 1/4 decrease with decreasing frequency at the high frequency range, but at the low frequency range a plateau region is observed. PMPI 1/4, having higher gel content, has a constant storage modulus over the entire frequency range measured, an indication of a crosslinked structure. Amine-extended

PP-g-MA with a molar ratio of 1/2 also has a plateau modulus.

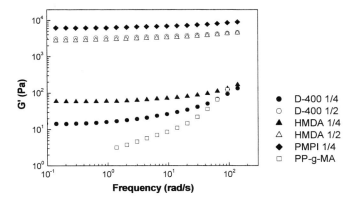

Figure 4. Storage moduli versus frequency at 270^0C for chain-extended PP-g-MA and at 170^0C for PP-g-MA.

Conclusions

PP-g-MA was chain-extended by reactive melt processing using polymeric isocyanate and a primary diamine. Isocyanate only reacted with the carboxylic acid form of MA and amine reacted with both the carboxylic acid and anhydride form of MA. With increasing initial amount of chain extender, gel content and storage modulus increased. Though PMPI only reacted with carboxylic acid, its high functionality and localized reaction lead to high gel content and high storage modulus.

[1] B. J. Scheve, J. W. Mayfield, Jr., A. J. DeNicola, U.S.P 4,916,198, Himont, **1990.**
[2] X. Wang, C. Tzoganakis, G. L. Rempel, *J. Appl. Polym. Sci.* **1996**, *61*, 1395.
[3] M. Y. Ju, F. C. Chang, *Polymer* **2000**, *41*, 1719.
[4] Z. Song, W. E. Baker, *J. Polym. Sci.: Part A: Polym. Chem.* **1992**, *30*, 1589.
[5] F. P. Tseng, J. J. Lin, C. R. Tseng, F. C. Chang, *Polymer* **2001**, *42*, 713.

Macromol. Symp. **2004**, *214*, 299-306

New Approaches for the Development of Highly Stable Polypropylene

Minoru Terano,[1] *Boping Liu,*[1] *Hisayuki Nakatani*[2]

[1] School of Materials Science, Japan Advanced Institute of Science and Technology, 1-1, Asahidai, Tatsunokuchi, Ishikawa, 923-1292, Japan

[2] Center for Nano Materials and Technology, Japan Advanced Institute of Science and Technology, 1-1, Asahidai, Tatsunokuchi, Ishikawa, 923-1292, Japan

Summary: Effects of primary structure on the degradation of polypropylene (PP) were studied using PPs having a variety of tacticities or a random ethylene sequence with the aim of improving the stability of PP. Thermal stabilities of isotactic, atactic and syndiotactic PPs (iPP, aPP and sPP) having similar molecular weights were investigated in terms of stereoregularity under the same conditions. The sPP showed the highest stability, suggesting that the racemic structure of the main chain interferes with the chain reaction of radical species. Moreover, iPP was prepared to evaluate the thermal stability with a random ethylene sequence (ethylene-propylene random copolymer: rPP). The rPP was more stable than iPP. The existence of the ethylene sequence in the main chain was found to be effective in achieving high stability.

Keywords: degradation; polypropylene; primary structure; stereoregularity

Introduction

PP has been widely used for commercial products in the form of fibers and films. However, PP is known to be very vulnerable to oxidative degradation under the influence of elevated temperature and sunlight.[1-8] The PP degradation chemistry has been very extensively studied and has long been recognized as a free-radical chain reaction,[9] which leads to chain scission. It is generally accepted that this scission is responsible for degradation in polymer mechanical properties. The addition of stabilizers has been widely used to depress the radical reaction. However, it is difficult to maintain the performance of stabilizers over a long term because of the volitality.[10-12] In order to suppress the radical reaction during long-term use, the modification of iPP itself may be effective.

DOI: 10.1002/masy.200451022

In this study, the thermal degradation of PP was investigated in terms of its primary structure, such as stereoregularity and comonomer incorporation. The key concept of this work is to investigate the influence of primary structure by choosing suitable PP samples and conditions to be able to eliminate other factors. A better understanding of the basic aspects of the decomposition of PP with regard to its primary structure is indispensable for the further development of this area.

Effects of Stereoregularity on Thermal Degradation of PP

The degradation behavior of PP depends on several factors, such as external stimulation, chemical structure of the sample and impurities in the sample. In this study, our attention was focused on stereoregularity. With the aim of clarifying the contribution of the independent factors, the emphasis fell on the choice of suitable PP samples and appropriate conditions to avoid the superimposition of each factor. The resulting information is considered to provide more profound insight into the stability of PPs as well as to contribute to the elucidation of the explicit mechanism of the thermal degradation of PP.

Several criteria for evaluating thermal stability have been proposed.[4] Among them, the most important one is the temperature at which the molecular weight decreases during a certain period. The typical molding temperature in the industrial iPP processing is about 180 °C or higher.[13] Under this condition, the effects of the crystallinity and crystal structure of the sample could be negligible because the applied temperature was higher than the T_ms of PPs (shown in Table I). The effect of stereoregularity on the degradation of PP was investigated in the molten state.

Table I Characteristics of PPs

Sample	$\overline{M}n$	$\overline{M}w / \overline{M}n$	mmmm (%)	Tm (°C)
iPP	50,000	5.0	99	160
aPP	48,000	3.8	23	
sPP	45,000	1.6	>1 (92)[a]	150

a) rrrr = 92%

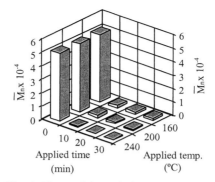

Fig. 1: Thermal degradation of iPP in air.

Recent catalyst technology makes it possible to produce suitable PP samples for the investigation. The dicyclopentadienyl zirconium dichloride-methylaluminoxane (MAO) catalyst system developed by Sinn and Kaminsky was used to produce aPP with high molecular weight.[14, 15] A syndiospecific metallocene catalyst, *i.e.*, isopropyl (cyclopentadienyl-1-fluorenyl) zirconium dichloride-MAO was successfully developed by Ewen[16] to make sPP with very high syndiotacticity (*rrrr* pentad fraction > 90%). iPP prepared by a MgCl$_2$-supported Ziegler catalyst system and aPP and sPP prepared by metallocene catalyst systems were used in this study.

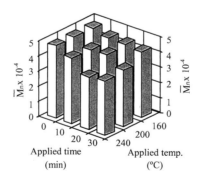

Fig. 2: Thermal degradation of aPP in air.

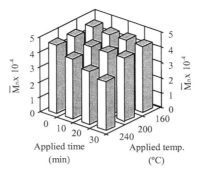

Fig. 3: Thermal degradation of sPP in air.

The molecular weights of these PPs were almost the same as shown in Table I. These PPs were used after the washing treatment. Heat treatments of PPs were carried out at 160-240 °C in air. Hence, the influence of molecular weight, catalyst residues, and crystalline state on the thermal degradation are negligible. The comparison of the thermal degradation of iPP, aPP and sPP was performed in air at 160, 200, and 240 °C. The degradation of iPP was found to proceed significantly (Fig. 1), whereas the degradation behavior was not remarkable in the case of aPP and sPP (Fig. 2 and 3). As can be seen in Fig. 4 (a) and (b), shifts in the GPC profile of the iPP and aPP to lower molecular weights were clearly observed without significant change of shape of the curves and molecular weight distribution after the heat treatment. In iPP, methyl groups are located on one side of the plane of the carbon-to-carbon main chain, and most of the configurational repeating units show meso-dyads. Many racemic structures exist in aPP. It seems reasonable to suppose that the thermal degradation is depressed significantly due to the existence of the racemic structure in aPP. In order to confirm this assumption, the thermal stability of sPP, which is mainly composed of the racemic structure, was investigated under the same conditions. As shown in Fig. 3, there was no significant difference in the $\overline{M}_n$ of the sPP before and after the heat treatments, during which iPP was degraded drastically. Furthermore, the shape of the GPC curve of sPP treated at 200 °C for 30 min is almost the same as that of the virgin sPP, as shown in Fig. 4(c). Thus, it was found that the thermal stability of sPP was appreciably higher than either that of iPP and/or aPP. It is inferred that good thermal stability of the sPP is mainly due to the presence of the racemic structure in the main chain.

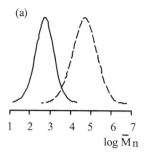

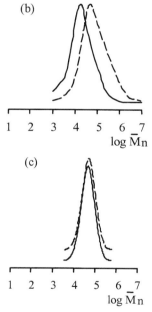

Fig. 4: GPC curves of iPP (a), aPP (b) and sPP (c). The solid lines indicate the results of treated PPS at 200°C for 30 min. The dotted lines indicate the results of virgin PPs.

On the basis of the results obtained, a feasible reason for the difference in the stabilities between iPP and sPP is considered to be as follows. Since the tertiary hydrogen in the sPP is located on the opposite side of the adjacent tertiary hydrogen, the scission of the main chain may hardly proceed even if the radical formation occurs on a tertiary hydrogen. In other words, this indicates that the stereoregularity is a dominant factor in PP degradation due to the reactivity of the peroxide radical. The effect of stereoregularity on thermal degradation of PP was also investigated by means of chemiluminescence, in which the difference in the signals between iPP and sPP was observed.[5]

Effect of Incorporation of Ethylene on Thermal Stability of iPP

Since the introduction of ethylene, which has no tertiary carbon atoms, effectively removes the reactive tertiary carbon atoms from the PP backbone, the modification of the primary structure of iPP by copolymerization with ethylene also seems to be an effective method for improving thermal stability. Therefore, the thermal stability of iPP and ethylene-propylene random copolymer (rPP) with a low ethylene content was investigated as a function of their primary structure.

The characteristics of iPP and rPP are summarized in Table II. In this study, very close molecular weight and molecular weight distribution of iPP and rPP were used.

Table II. Characteristics of iPP and rPP

Sample	$\overline{M}$	$\overline{M}w / \overline{M}$	mmmm	Et cont.[a] mol
iP	48,000	5.7	99	—
rP	41,00	6.	93	5.3

a) Ethylene content in polypropylene

Fig. 5 showed the heat treatment time dependence of the GPC curves of iPP and rPP. The curve shift between the virgin PP and the degraded one was shown. The curves of both PPs were shifted towards lower molecular weight with the heat treatment time. However, the rate of decrease in the molecular weight of rPP is considerably slower than that of iPP, implying that rPP has a higher thermal stability.

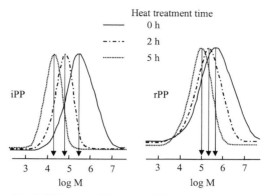

Fig. 5: GPC curves of heat treated PPs at 180 °C in air.

The PP degradation chemistry is believed to propagate by a free radical chain reaction. The formation of alkyl radical by the dissociation of tertiary C-H bond firstly occurs. The reaction of this alkyl radical with oxygen leads to the production of the peroxy radical, which propagates the chain reaction by intramolecular tertiary C-H abstraction from iPP as well as secondary C-H abstraction from rPP. The abstraction reactions produce the hydroperoxides which are responsible for the autocatalytic nature of the degradation. The existence of the secondary C-H in rPP seems to depress the chain reaction of the hydroperoxide regardless of its low content of ethylene.[8]

Conclusions

In this study, the thermal degradation behavior of PPs with a variety of tacticities and ethylene incorporation was investigated as a function of the primary structure in order to improve the stability of PP. The sPP and rPP showed better stability compared to iPP.

The stability of PP was found to be greatly improved by the modification of the primary structure, such as stereoregularity and comonomer incorporation.

306

[1]   Y. Kato, D. J. Carlsson, D. M. Wiles, *J Appl. Polym. Sci.* **1969,** *13,* 1447.
[2]   J. B. Knight, P. D. Calvert, N. C. Billingham, *Polymer* **1985,** *26,* 1713.
[3]   G. A. George, M. Celina, A. M. Vassallo, P. A. Cole-Clarke, *Polym. Degrad. Stab.* **1995,** *48,* 199.
[4]   H. Mori, T. Hatanaka, M. Terano, *Macromol. Rapid. Commun.* **1997,** *18,* 157.
[5]   Z. Osawa, M. Kato, M. Terano,. *Macromol. Rapid. Commun.* **1997,** *18,* 667.
[6]   T. Hatanaka, H. Mori, M. Terano, *Polym. Degrad. Stab.* **1999,** *64,* 313
[7]   H. Nakatani, T. Hatanaka, K. Nitta, M. Terano, *Res. Adv. in Macromolecules* **2000,** *1,* 17.
[8]   M. S. Alam, H. Nakatani, T. Ichiki, G. S. Goss, B. Liu, M. Terano, *J. Appl. Polym. Sci.* **2002,** *86,*1863.
[9]   J. C. W. Chien, D. S. T. Wang, *Macromolecules* **1975,** *8,* 920.
[10]   R. Gensler, C. J. G. Plummer, H. H. Kausch, E. Kramer, J. R. Pauquet, H. Zweifel, *Polym. Degrad. Stab.* **2000,** *67,* 195.
[11]   P. P. Klemchuk, P. L. Horng, *Polym. Degrad. Stab.* **1991,** *34,* 333.
[12]   J. Pospisil, *Polym. Degrad. Stab.* **1993,** *34,* 230.
[13]   See books on molding of polypropylene, e.g., S. S. Petrovan, *"Handbook of Polyolefins"*, Marcel Dekker, New York, 1993.
[14]   H. Sinn and W. Kaminsky, *Adv. Organomet. Chem.* **1980,** *18,* 99.
[15]   W. Kaminsky, *Macromol. Chem. Phys.* **1996,** *197,* 3907.
[16]   J. A. Ewen, R. L. Jones, and A. Razavi, *J. Am. Chem. Soc.* **1998,** *110,* 6255.

Structure, Dynamics and Properties of Materials with Polymers Having Complex Architectures

Tadeusz Pakula

Max-Planck-Institute for Polymer Research, Postfach 3148, 55021 Mainz, Germany
E-mail: pakula@mpip-mainz.mpg.de

Summary: Various non-linear highly branched polymers such as multiarm stars, block copolymer micelles and bottlebrush-like polymers have been studied in order to analyze their intramolecular structure and effects of spatial ordering resulting from their specific macromolecular architecture. These polymers constitute a class of compact macromolecules which, due to the high intramolecular density, interact strongly, excluding each other in space. Investigations of the structure and dynamics in such systems, using various experimental methods as well as computer simulations, have been performed. Small angle X-ray scattering is used to characterize the structure and mechanical spectroscopy is used for the detection of the dynamic behavior of the systems. Simulations have been performed using the cooperative motion algorithm with lattice polymers equivalent to the considered macromolecules. Results of both experiments and the simulation seem to support the concept of slow structural cooperative rearrangements controlling the flow in such systems. The effects are analogous to those related to flow in low molecular liquids but take place on another size scale. The new slow relaxation processes creates new super soft-states which are characterized by shear modulus plateau lower than 10^4 Pa.

Keywords: complex polymer architectures; dynamics; relaxation processes; soft materials; structured melts

Introduction

New complex macromolecular architectures are of particular interest when they lead to new material properties. A large variety of macromolecules differing by topology of bond skeletons and by distributions of monomers of different types have been synthesized [1-4]. In many cases, such macromolecules constitute self-assembling systems in which a supramolecular order can result in modified dynamics and, consequently, in new properties [5,6]. In the simplest case of linear polymers, joining monomeric units into linear chains results in a dramatic change of properties. Whereas a system consisting of monomers can usually be only liquid-like or solid (e.g. glassy), the polymer can additionally exhibit a rubbery

DOI: 10.1002/masy.200451023

state, with properties that make these materials extraordinary in a large number of applications. This new state is manifested by the very slow relaxation of polymer chains in comparison with the fast motion of monomers, especially when the chains become so long that they can entangle.

Knowledge of the mechanical behavior for complex macromolecular architectures is not as comprehensive as for the linear polymers. In this paper, some effects of macromolecular architecture on mechanical properties will therefore be illustrated and discussed. The structures considered are illustrated in Figure 1.

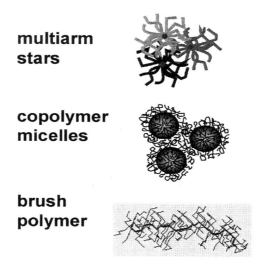

multiarm stars

copolymer micelles

brush polymer

Figure 1. Illustration of the macromolecular architectures considered in this paper.

Experimental Results

Mechanical manifestation of relaxations taking place in a linear polymer melt in comparison with the behavior of a low molecular system is illustrated in Figure 2. The dynamic mechanical characteristics of the two materials indicate a single relaxation in the monomer system in contrast to the two characteristic relaxations in the polymer. The rubbery state, characteristic of entangled polymers, extends between the segmental (monomer) and the chain relaxation frequencies and is controlled by the chain length determining the ratio of the two relaxation rates. In the rubbery state the material is much softer than in the solid state. If expressed by the real part of the modulus, the typical solid-state elasticity is of the order of 10^9 Pa or higher, whereas the rubber like elasticity in bulk polymers is of the order of 10^5-10^6 Pa.

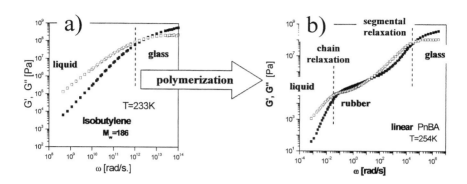

Figure 2.   Real (G') and imaginary (G'') parts of the complex shear modulus vs. frequency for (a) an isobutylene oligomer and (b) a melt of linear p(n-butyl acrylate), p(nBA), chains.

An example of the behavior of multiarm stars in the melt is illustrated in Figure 3, where the mechanical results indicate three relaxation processes. The slowest is related to rearrangements within the supramolecular order, documented by means of the small angle X-ray scattering result presented in the insert [7].

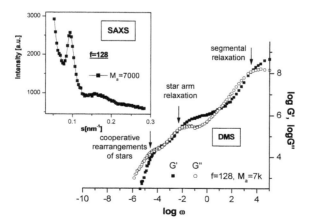

Figure 3.   An example of the small angle X-ray scattering (SAXS) and dynamic mechanical spectroscopy (DMS) results for a polybutadiene melt of multiarm stars. The scattering peak indicating star ordering and various modes in the mechanical relaxation can be seen.

It is interesting to notice that there is a new dynamic state in this system between the relaxation of the arms and the structural relaxation. In this state, the melt of multiarm stars shows a new elastic plateau with the G' level which is considerably lower than that characteristic of the entangled linear polymers.

Another example of the same type of behavior is illustrated in Figure 4. In this case, spherical micelles of polyisoprene-b-polystyrene block copolymers dispersed in a polyisoprene matrix show ordering, which depends on the micel concentration, as illustrated in Figure 4c. The dynamic mechanical results indicate that with the increasing order a new slow relaxation process is created which, also here, can be attributed to structural relaxation [8]. For the system containing 50% micelles this structural relaxation becomes so slow that the G' plateau at the level of 10^3 Pa extends to the lowest detectable frequencies.

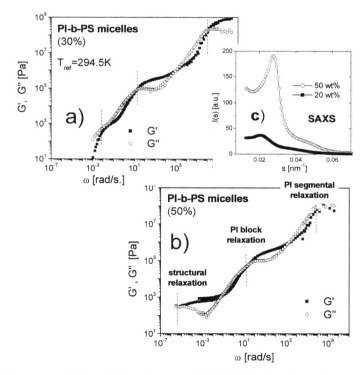

Figure 4. Master curves of G' and G'' vs. frequency for PI-b-PS spherical micelles dispersed in PI matrix with concentration (a) 30% and (b) 50%. Vertical dashed lines indicate the relaxation processes as assigned in (b). In (c) the SAXS intensity distributions for the two systems are shown.

In the example given in Figure 5, viscoelastic properties determined for the melt of brush-like macromolecules are presented as master curves of G' and G'' determined for the reference temperature, 254 K [9]. The results indicate a presence of three relaxation ranges: the high frequency relaxation corresponds to segmental motion, the intermediate relaxation is attributed to the reorientation rates of the side chains and the slowest process is the global macromolecular relaxation in this system, which controls the zero shear flow and the corresponding viscosity. The rate and nature of this relaxation must be dependent on the length of the backbone. For short backbone chains these macromolecules can behave similarly to stars, for which translational motion dominates the slow dynamics, but for longer backbones, the reorientation possibilities should become slower than translation [10].

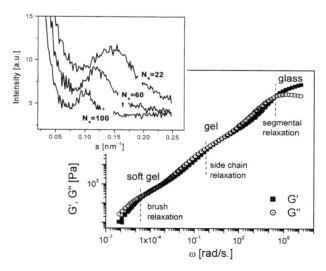

Figure 5. Frequency dependencies of the storage (G') and loss (G'') shear moduli (master curves at T_{ref}=254 K). The insert: Small angle X-ray scattering from melts of p(nBA) brush like polymer with various side chain length.

Discussion

Complex polymeric systems such as the discussed melts of multiarm stars or melts of micelles in microphase separated block copolymers seem to show a complex dynamic behavior resulting from the ordering of stars or micelles. The results presented indicated that in both these cases three relaxation processes are detected, the slowest one of which must be related to the structural rearrangements. Dense systems of stars considered as examples of macromolecules of this kind have been simulated using the cooperative motion algorithm [7,11]. Results have shown that the structure in such systems develops due to strong excluded volume effect on the macromolecular scale, because the multiarm stars (with number of arms greater than 25) fill locally the space, becoming impenetrable. Development of the structure leads to additional dynamic slow relaxation modes, which can be considered as having similar origin for both stars and hairy micelles. The flow of such systems becomes controlled not by the relaxation of polymeric arms but by the additional slow relaxation process, which is attributed to cooperative rearrangements of the larger elements (stars or micelles) within the ordered state. This leads to a slowing down of the decay of the position correlation of the stars or micelles which, for a large number of polymeric arms, becomes considerably slower than the relaxation of the arms and constitutes in this way an extra slow mode in the translational relaxation. This relaxation, although taking place on the macromolecular size scale, seems to have many similarities with the cooperative rearrangements considered in the dynamics of small molecules in the simple liquid [12]. The degree of order in the studied systems is mainly controlled by the ratio of the core radius to corona thickness which evidently depends on various molecular parameters, such as length and number of arms. For stars or miceles with a large number of short arms, the highest degree of ordering should be expected, whereas, stars or micelles with long arms could show a limited order even when the number of arms is high. These effects have been observed both in real and in simulated systems [7,8,13].

The type of order observed in the multiarm star melts can be described as liquid-like on the macromolecular scale. Neither in real nor in simulated systems have any clear signatures of lattice formation been detected. This probably results from the deformability and related form fluctuations of the star coronas. They consist of flexible arms and remain soft spheres even when the number of arms is large and the core radius is large. The latter is well confirmed both by the dynamic properties of the stars and in polymer melt filled with spherical copolymer micelles.

Qualitatively, the simulation results concerning dynamics seem to be in good agreement with the experimental observations. Both the relaxations detected by means of the viscoelastic measurements and the suggested assignments are reflected in the simulation. Two relaxations, one related to segmental motion and the other to arm relaxation, have been observed in all systems. The relaxation rate of the first one is independent of the structure parameters both in experiments and in the simulation. On the other hand, the arm relaxation has been observed to be considerably dependent on the arm length but essentially independent of the arm number. The most interesting effect observed both in the simulated and in the real systems was the additional slow relaxation process appearing in systems with clear ordering. The analysis of the simulation results concerning dynamics, as well as a direct observation of star motions in these systems (Fig. 6), leads to the conjecture that the slow process can be related to translational cooperative rearrangements of stars within the ordered state which are of the same character as those suggested for the rearrangements in low molecular liquids. This might be a cooperative process on the macromolecular scale with the mechanism analogous to that postulated for liquids in the dynamic lattice liquid model [12,14]. This model postulates that the rearrangements of macromolecules take place along closed loops as a result of preservation of local continuity in the dense ordered system.

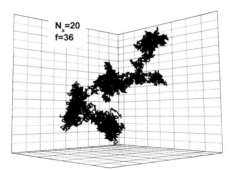

Fig. 6. Typical star center of mass trajectory recorded for a multiarm star in the simulated melt. The trajectory consists of blobs related to a longer residence of the star at some well distinguishable places which are distributed along the trajectory and are connected by other trajectory fragments related to faster displacements between the localized states.

This mechanism of relaxation can be considered for stars, hairy micelles and probably for short brush macromolecules which are not much different from the stars. For longer brush macromolecules, the orientational relaxation should become the slowest and for the brushes with the backbone sufficiently long to form entanglements the entanglement, relaxation should become dominant.

In all the studied systems, the intermediate relaxation which is attributed to motion of arms or hairs, plays an important and specific role. If the arms or "hairs" are short, this relaxation is responsible for a drop of modulus to a level considerably lower than that characteristic for conventional polymers. The states between the longest relaxation and the arm relaxation constitute a possibility to create new mechanical properties, which can be classified as super-soft [15]. Depending on structural parameters of the systems, the proportions between relaxation rates corresponding to three distinguished processes can be influenced. This should lead to various properties of materials in frequency ranges corresponding to modulus plateaus related to various unrelaxed states i.e. glassy, rubbery or the super-soft state. For example, extension of side chain length to the range where they can entangle should extend the frequency range of the modulus plateau at the typical level for polymer rubber elasticity, i.e. at 10^5-10^6 Pa (in detail, depending on the chemical nature of the monomer). On the other hand, extension of the backbone with not yet entangled side chains can lead to an extension of the frequency range with the super-soft elastic plateau with a modulus level below 10^4 Pa [15].

Conclusions

It is pointed out that some polymers with complex architectures represent model soft objects, which order on a macromolecular scale because of strong steric interactions. This is a consequence of their complex topology, which leads to specific intramolecular monomer density distributions. It is shown that signatures of the structures formed appear in the dynamics of these systems as extra slow relaxation processes, which can control the terminal rheological behavior as well as lead to new properties of unusually soft bulk materials. The simulation results are able to represent the structure and dynamics of the considered systems.

Acknowledgement

The author is grateful to K. Gohr, G. Lindenblatt, K. Matyjaszewski, P. Minkin, W. Schärtl, M. Schmidt and D. Vlassopoulos for their fruitfull cooperation in the field discussed in this paper.

[1]   D. B. Ambilino, J. F. Stoddart, *Chem. Rev.* **1995**, *95*, 2725.
[2]   L. J. Fetters, A. D.Kiss, D. S. Pearson, G. F. Quack, F. J. Vitus, *Macromolecules* **1993**, *26*, 647.
[3]   J. Roovers, Zbou Lin-Lin, P. M. Toporowski, M. van der Zwan, H. Iatron, N. Hadjichristidis, *Macromolecules* **1993**, *25*, 4324.
[4]   K. Matyjaszewski, *"Advances in Controlled/Living Radical Polymerization"*, ACS Symp. Series 854, Washington, DC, 2003.

[5]   S. Ch.Hong, T. Pakula, K. Matyjaszewski, *Macromol.Chem. Phys.* **2001**, *202*, 3392.

[6]   H. Shinoda, K. Matyjaszewski, L. Okrasa, M. Mierzwa, T. Pakula, *Macromolecules* **2003**, *36*, 4772.

[7]   T. Pakula, D. Vlassopoulos, G. Fytas, J. Roovers, *Macromolecules* **1998**, *31*, 8931.

[8]   K. Gohr, T. Pakula, K. Tsutsumi, W. Schärtl, *Macromolecules* **1999**, *32*, 7156.

[9]   T. Pakula, P. Minkin, K. L. Beers, K. Matyjaszewski, Abs. Pap. ACS 2001, Vol 221, pp 559-PMSE.

[10]   T. Pakula, K. Harre, *Comput. Theor. Polym. Sci.* **2000**, *10*, 197.

[11]   T. Pakula, *Comput. Theor. Polym. Sci.* **1998**, *8*, 21.

[12]   T. Pakula, J. Teichmann, *Mat. Res. Soc. Symp. Proc.* **1997**, *45*, 211.

[13]   G. Lindenblatt, W. Schartl, T. Pakula, M. Schmidt, *Macromolecules* **2001**, *34*, 1730.

[14]   T. Pakula, *J. Mol. Liquids* **2000**, *86*, 109.

[15]   D. Neugebauer, Y. Zhang, T. Pakula, S. S. Sheiko, K. Matyjaszewski, *Macromolecules* – in press.

Morphology and Temperature Phase Transitions in α,ω-Alkanediols with Different Chain Lengths

Vyacheslav Marikhin,[*1] *Victor Egorov,*[1] *Elena Ivan'kova,*[1] *Liubov Myasnikova,*[1]

Elena Radovanova,[1] *Boris Volchek,*[2] *Darya Medvedeva,*[2] *Alan Jonas*[3]

[1]Ioffe Physico-Technical Institute of RAS, Polytekhnicheskaya 26, St. Petersburg 194021, Russia
E-mail: v.marikhin@mail.ioffe.ru
[2]Institute of Macromolecular Compounds of RAS, Bolshoy pr. V.O. 31, St. Petersburg 199004, Russia
[3]Universite Catholique de Louvain, Place Croix du Sud, 1B-1348 Louvain-la-Neuve, Belgium

Summary: A comparative analysis of phase transitions in diols with various chain lengths $[(CH_2)_{44}(OH)_2$ and $(CH_2)_{22}(OH)_2]$ and changes in their absorption spectra with temperature have been investigated by DSC and FTIR. Analysis of the DSC data has led to the conclusion that the low-temperature phase transition of $(CH_2)_{22}(OH)_2$ in a solid state (T_{s-s} = 367.1 K) is a phase transition of the first order, while the high-temperature phase transition (T_m = 376.3 K) is of the second order, i.e., a transition of the order-disorder type. Splitting of the IR absorption bands into doublets at 720-730 cm^{-1} and 1463-1473 cm^{-1} indicates that crystalline subcells in the lamellae of both diols are orthorhombic lattices with the parameters typical of hydrocarbons. IR spectra showed that at the phase transition temperature T_{s-s} transformation of an orthorhombic subcell into a hexagonal one occurs. This type of molecular chain packing remains the same up to the melting temperature T_m. In a $(CH_2)_{44}(OH)_2$ diol, the ortho-hexagonal subcell transition occurs only at the melting temperature (390.0 K). The wide IR band in the region from 3000 cm^{-1} to 3600 cm^{-1} shows that end hydroxyl groups of diol molecules form, on the surfaces of lamellar crystals, long (polymer) regular sequences consisting of intermolecular hydrogen bonds.

Keywords: α,ω-alkanediols; DSC; FT-IR; morphology; phase behavior

Introduction

Due to their monodisperse molecular mass and absence of chemical defects, long-chain molecular crystals are suitable models aimed at solving many of the controversial problems concerning the relationship between structure and physical and chemical properties of polymers crystals, such as the influence of chain length and types of end groups on structure formation during crystallization from solutions and melts, structural and conformational transformations at phase transitions, the nature and energy of generation of conformational and translational defects, etc. Particular advantages of long-chain molecular crystals over conventional polydisperse polymers with chemical defects are offered in establishing general quantitative relationships between structure and different properties of materials.

The simplest long-chain molecular crystals are n-paraffins, but even their structure and properties are rather complicated and depend on the number (n) of CH_2 groups in the methylene sequence. In addition, the properties of n-paraffins depend on whether n is odd or even[1-15]. The structure and properties of long-chain α,ω-alkanediols have not been extensively studied as the paraffins have.[16-23] The α,ω-alkanediols are particularly attractive because they allow elucidation of the influence of the type of end groups on the structure and properties of diol crystals in comparison with n-paraffins with the same chain length (the methylene parts of both compounds contain repeating CH_2 units). Interest in diols is also due to the fact that end hydroxyl groups on basal planes of lamellar crystals form long "polymer" chains consisting of hydrogen bonds. This leads to a significant increase in the surface energy of these crystals and, as a consequence, to changes in their thermodynamical characteristics, such as melting temperature, specific features of phase transitions in the heating-cooling cycles, etc.[16-23]

The goal of our work was to study the morphology of diols and its transformation at temperature phase transitions in diols which have the chain lengths differing by a factor of two, i.e., 1,22-docosanediol (D-22) and 1,44-tetratetracontanediol (D-44).

Experimental

Materials

The compounds studied were 1,22-docosanediol (D-22) and 1,44-tetratetracontanediol (D-44), synthesized in the laboratory of Prof. A. Jonas (Belgium) from commercially available reagents. D-22 was synthesized using a procedure that was a slight modification of that described in [23]. D-22 and was recrystallized from methanol. D-44 was synthesized from D-22 by using the same improved procedure as for D-22. The final product was purified by recrystallization from benzene and discoloration with silica.

Instrumentation

Morphology of the samples was investigated by a JEOL-6300 scanning microscope. A Perkin-Elmer DSC-2 was used for calorimetry studies. Conventional procedures were employed for device calibration. The velocity of scanning V of the thermal flow was varied from V = 1.25 K·min^{-1} to V = 10 K·min^{-1}. IR absorption spectra, from room temperature to the melting temperatures of the samples, were recorded by a Bruker JFS-88 (Bruker Analytik GmbH, Pheinstetten, Germany) Fourier spectrometer over a wide range of frequencies, from 500 cm^{-1} to 5000 cm^{-1}.

Results and Discussion

Scanning Electron Microscopy (SEM)

Figures 1 and 2 show scanning electron micrographs of morphological units of the lamellar type arising in the products of synthesis during polymerization-crystallization.

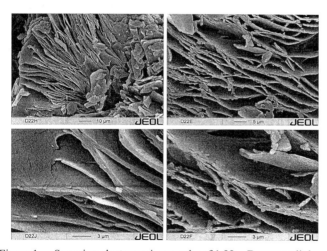

Figure 1.   Scanning electron micrographs of 1,22 – Docosanediol.

It can be seen that both types of diols are characterized by formation of fairly long (tens of microns) structures of the lamellar type whose thickness can be as large as units of microns. A tendency toward formation of spherulite-like units is more pronounced in D-22 than in D-44 (Figure 1).

It is known[24-29] that long-chain molecules in molecular crystals form single lamellae from fully extended chains whose monodisperse thickness is proportional to the molecule length, i.e., it amounts to several tens of nanometers. Therefore, it can be concluded that lamellar structures seen on SEM micrographs (Figures 1 and 2) consist of stacks of several hundreds-thousands of single-crystal lamellae. It has been shown[17, 18] that in diols with an even number of CH_2 groups the chains are tilted at an angle (30-35°) to the normal to the basal planes formed by end hydroxyl groups. In addition, when lamellae are superposed on each other, the tilts of the angles in neighboring lamellae change to the opposite ones, and, as a consequence, structures of the herring-bone motif arise.[18, 19]

Figure 2.   Scanning electron micrographs of 1,44 – Tetratetracontanediol.

Differential Scanning Calorimetry (DSC)

It has been found[16,19] that in diols, like in n-paraffins, phase transitions are observed, even in the solid state. Their number depends on the chain length and on whether the number of carbon atoms in the molecule backbone is odd or even.

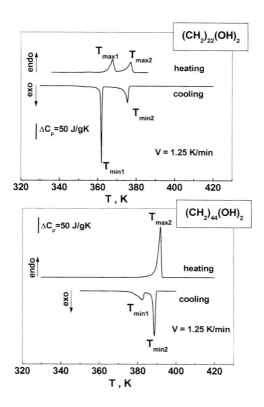

Figure 3.   Heating – cooling DSC thermographs of α,ω-alkanediols.

Figure 3 presents the DSC thermograms obtained for the D-22 and D-44 samples with a mass of about 3 mg in the heating-cooling regime. The scanning velocity was V = 1.25 K·min[-1]. It is evident from Figure 3 that the thermogram of the D-22 sample heated from room temperature exhibits an endothermal maximum at T_{max1} (about 367 K), which is attributable to a phase transition in the solid state T_{s-s}. At T_{max2} (in the vicinity of 377 K), melting of the sample takes place (order-disorder transition). When the sample is cooled from the melt, it crystallizes at T_{min2} (disorder-order transition). Further cooling is accompanied by a pronounced exothermal effect, attributable to the phase transition in the solid state at T_{min1}. In contrast to D-22, when D-44 is heated, transition from the low-temperature crystalline state into the melt occurs in one stage at the melting temperature T_{max2} (in the vicinity of 392 K). No transition of the T_{s-s} type was observed for this sample in the heating cycle. On the other hand, for the D-44 sample cooled from the melt, the second endothermal peak at T_{min1} attributable to phase transition in the solid state is observed in addition to the exothermal peak at T_{min2} caused by crystallization. On the whole, the observed picture agrees with the literature data[16,19]. Displacements of relative positions of maxima of endo- and exo-peaks during the heating-cooling process (the hysteresis effect) can be an important indication of the nature of phase transitions[30]. To be more exact, the effects of hysteresis should be observed for the transitions of the first order (structural transitions), while for the transitions of the second order they must be absent.

However[30], it is known that experimental errors (displacements of peaks) associated with the effect of thermo resistance, that strongly depends on the mass of the sample and scanning velocity, can occur in DSC studies.

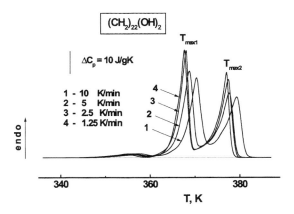

Figure 4.   DSC thermographs of 1,22 – Docosanediol recorded at various scan rate.

As an example, Figure 4 shows thermograms of the D-22 sample obtained when the heating rate was varied from 1.25 K·min^{-1} to 10 K·min^{-1}. It can be seen that not only do the positions of the peaks at T_{max1} and T_{max2} shift by several degrees, but also the important characteristic of the half widths of the maxima changes considerably.

Illers showed[31] that experimental errors can be eliminated. To this end, from the experimental data obtained for a wide range of heating rates V, the $T_{max(exp)} = f(V^{1/2})$ dependencies, which must be linear in the case of absence of any phase transitions, are plotted. Extrapolation of the linear dependencies to $V \rightarrow 0$ yields the true transition temperatures.

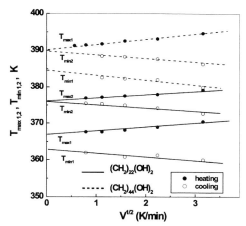

Figure 5. Dependence of phase transition temperatures on scan rate for 1,22 – Docosanediol and 1,44 – Tetratetracontanediol.

Figure 5 shows the $T_{max(exp)} = f(V^{1/2})$ dependencies for the D-22 and D-44 samples in the heating-cooling cycles. Table 1 gives extrapolated (true) phase transition temperatures T_{max1} and T_{min1}, and also T_{max2} and T_{min2}.

The data given in Figure 5 and Table 1 indicate that there is no hysteresis for positions of second maxima for D-22 and D-44 in the heating-cooling cycles. In accordance with the theories of phase transition,[30] these DSC data lead to the conclusion that the high-temperature transition in the diols studied is indeed the second-order transition of the order-disorder type, and the hysteresis observed earlier in a number of studies in heating-cooling cycles was due to experimental inaccuracies.

Table 1.   True thermodynamical parameters of phase transition in 1,22 – Docosanediol and 1,44 – Teteratetracosanediol.

Sample	T_{max1}^{r}	T_{max2}^{r}	T_{min1}^{r}	T_{min2}^{r}	ΔT_1	ΔT_2	ΔH_1	ΔH_2	ΔS_{exp} 1	ΔS_{exp} 2
	K	K	K	K	K	K	J/g	J/g	J/gK	J/gK
$C_{22}(OH)_2$	367.1	376.3	362.8	376.3	4.3	0	108.6	151.4	–	0.407
$C_{44}(OH)_2$	–	390.0	384.0	390.0	6.0	0	–	288.7	–	0.740

On the other hand, a pronounced hysteresis is observed for the low-temperature transition, even for extrapolated temperatures at $V \to 0$, which can be an additional argument showing that this transition is of the first order, i.e., the structural transition associated typically with the change of the crystallographic subcell type (see below).

Additional information on the nature of observed phase transitions can also be obtained from analysis of the shape of peaks on the thermograms of the samples studied.

It has been shown[32] that, for the first-order transition, the endothermal peak must be δ-like shaped and have hysteresis in the heating-cooling cycle.

On the other hand, characteristic features of the transition of the order-disorder type (of the second order) is a λ-like shape of the $C_p(T)$ peak and the absence of hysteresis in the heating-cooling cycle.[33] Figure 6 demonstrates for D-22, as an example, that these requirements for the peak shapes are fulfilled when each of the maxima (Figure 6b and c) is separated out from the complicated thermogram (Figure 6a).

It has also been shown[30] that at temperatures $T < T_{max2}$ the $C_p(T)$ dependence is described by the power function of the type

$$_p(T) = A + B(T_0 - T)^{-n}, \tag{1}$$

where A, B, T_0, and n are constants. T_0 has the meaning of the true temperature of the second-order phase transition.

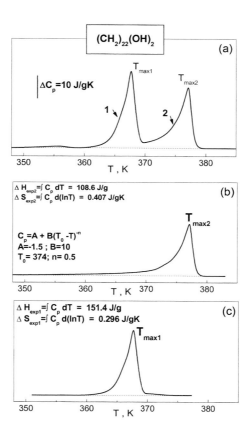

Figure 6. Deconvolution of experimental DSC endo thermogram of 1,22 – Docosanediol (a) into two peaks corresponding T_{max2} – order-disorder transition (b) and T_{max1} – solid-solid transition T_{ss} (c).

For the experimental $C_p(T)$ dependence of D-22, a good agreement between calculated and experimental curves was obtained if fitting parameters A = 1.5, B = 10, n = 0.5, and $T_0 = 374$ K were used.

Note that the obtained $T_0 = 374$ K differs from the value given in Table 1 ($T_{max2} = 376.3$ K). The reasons for this discrepancy are unclear at present.

It is believed[30, 33, 34] that the abnormal change in heat capacity at the order-disorder transition is due to a change in the conformational entropy, whose limiting value at the crystal-melt transition is[34]

$$S_c = R \cdot \ln 2 = 5.73 \text{ J} \cdot \text{mol}^{-1} \cdot \text{K}^{-1} \tag{2}$$

The experimental value of ΔS_c determined from

$$\Delta S_{exp}^{max 2} = \int C_p(T) d(\ln T) \tag{3}$$

turned out to be

$$\Delta S_{exp}^{max 2} = 0.407 \text{ Jg}^{-1} \text{K}^{-1} \tag{4}$$

Comparison of Equation (2) and Equation (4) yields the value of "mole" in our case

$$\Delta S_{exp}^{max 2} = 0.407 \text{ Jg}^{-1} \text{K}^{-1} \equiv \Delta S_c = 5.73 \text{ J} \cdot \text{mol}^{-1} \text{K}^{-1}$$

i.e.,

$$1 \text{ mole} = 14.08 \text{ g},$$

which corresponds to the mol mass of the CH_2 group. This is consistent with the statement that the main role in realization of the second-order transition is played by kinetic units consisting of CH_2 groups.

Associated with this, it is interesting to find what contribution to this transition comes from end hydroxyl groups which, by forming a network of hydrogen bonds on the surfaces of elementary lamellae, prevent mobility under unfreezing. In particular, they considerably increase the melting temperature of diols compared with n-paraffins with the same number of carbon atoms in the chain.

Let us consider the entropy of a system consisting of two basic constituents - CH_2 groups and CH_2OH groups. For such a system, the entropy is

$$\Delta S_c = (n_1 + n_2)\, R \cdot \ln 2 = (N_1/M_d + N_2/M_d \cdot \frac{1}{k})R \cdot \ln 2 = 5.73\ \text{J} \cdot \text{mol}^{-1}\ \text{K}^{-1}.$$

where N_1 is the number of moles of CH_2 groups in one mole of diol ($N_1 = 20$ for D-22); N_2 is the number of moles of CH_2OH groups in one mole of diol ($N_2 = 2$ for D-22).

$$K = \frac{M_{CH_2}}{M_{CH_2OH}}\ ; \qquad M_{CH_2} = 14;\ M_{CH_2OH} = 31$$

$$M_{D-22} = 342;\quad K = \frac{14}{31} \cong 0.452$$

$$n_1 = \frac{N_1}{M_d} = \frac{20}{342} = 0.0585\ \text{mol/g}$$

$$n_2 = \frac{N_2}{M_d} \cdot \frac{1}{K} = \frac{2}{342 \cdot 0.452} = 0.0129\ \text{mol/g}$$

Then

$$\Delta S_c \cong (0.0585 + 0.0129) \cdot 5.37\ \text{J} \cdot \text{mol}^{-1}\ \text{K}^{-1} = 0.409\ \text{J} \cdot \text{mol}^{-1}\ \text{K}^{-1} \approx \Delta S_{exp2} = 0.407\ \text{J} \cdot \text{mol}^{-1}\ \text{K}^{-1}.$$

Thus, our model is in good agreement with experimental data and allows estimation of the relative contributions of CH_2 and CH_2OH groups into the entropy of the order-disorder transition as

$$W = 0.129/0.0585 \cong 0.22 \cong 22\ \%$$

These estimates show that despite a relatively low ($\sim$10 %) mole fraction of CH_2OH end groups in the molecule of D-22, their contribution to a change of entropy at the order-disorder transition

328

is fairly large, i.e., about 22 %.

As noted above, the low-temperature (structural) first-order transition is typically attributed to the change in the type of packing of molecules in subcells of crystalline lamellae accompanied by conversion of subcells with lower symmetry into subcells with symmetry of a higher order. In the literature on X-ray structural analysis of diols[17, 18], data on parameters of subcells are generally not, though it is supposed that packing of methylene sequences at low temperatures is orthorhombic and the subcell parameters of diols are similar to those of n-paraffins and polyethylene[35], i.e., $a_s = 7.4$ A, $b_s = 4.96$ A, and $c_s = 2.54$ A. It is also believed[36] that the first-order transition is accompanied by transformation of the orthorhombic elementary subcells into hexagonal (rotational) subcells.

Valuable information on the type of subcells and character of structural transformations at phase transitions in D-22 and D-44 can be obtained from analysis of IR spectroscopic data.

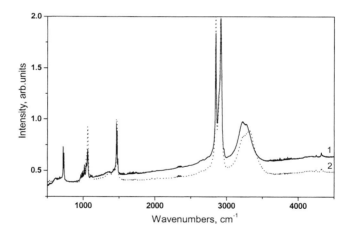

Figure 7.   FTIR spectra of 1,22 – Docosanediol (1) and 1,44 – Tetratetracontanediol (2).

Figure 7 shows IR Fourier spectra in the frequency range from 500 cm^{-1} to 4500 cm^{1} for D-22 (curve 1) and D-44 (curve 2). The spectra are seen to have intense characteristic absorption bands, typical of crystals consisting of methylene sequences with orthorhombic subcells, i.e., well known doublets of methylene rocking mode bands at 720-730 cm^{-1}, methylene bending mode bands at 1463-1473 cm^{-1}, and symmetric and asymmetric stretching C-H mode bands at $v_s = 2850$ cm^{-1} and $v_a = 2917$ cm^{-1}[37].

In addition to the bands caused by the methylene part of the molecules, the spectra exhibit a narrow band at v=1067 cm^{-1} and a very broad band in the region from 3000 cm^{-1} to 3600 cm^{-1}.

According to literature[38], the band at v = 1067 cm^{-1} corresponds to the C-O stretching mode of a hydroxyl group, and the broad band is due to the O-H stretching mode of the hydroxyl groups forming intermolecular polymer sequences consisting of hydrogen bonds of the (O-H...O) type (see below).

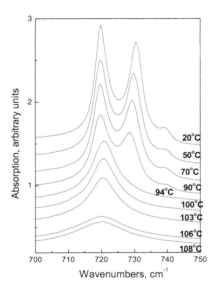

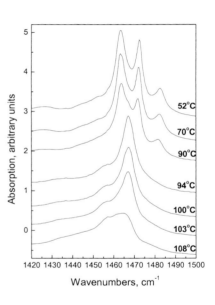

Figure 8. Temperature dependence of methylene rocking mode doublets in 1,22 – Docosanediol.

Figure 9. Temperature dependence of methylene bending mode doublets in 1,22 – Docosanediol.

Figures 8-11 show changes in the doublets at 720/730 cm⁻¹ and 1463/1473 cm⁻¹ as temperature is increased from room temperature to T > T$_m$ for D-22 and D-44. It is obvious from these figures that there are phase transformations in subcells.

The temperature changes in the doublets of diols are similar to the behavior of analogous doublets arising on heating of n-paraffins and polyethylene.

First of all, information on the change in the subcell type can be inferred from the behavior of the high-frequency components (730 cm⁻¹ and 1473 cm⁻¹) of doublets. It is known[39] that just the high-frequency component appears in IR spectra of hydrocarbons as factor group splitting resulting from specific intermolecular interactions between two chains in the orthorhombic subcell when the planes of trans-zigzags in neighboring layers of the molecules are perpendicular to each other. If this specific arrangement is disturbed (for example, in a hexagonal subcell), a single band ascribed to an isolated trans-zigzag is observed instead of a doublet.

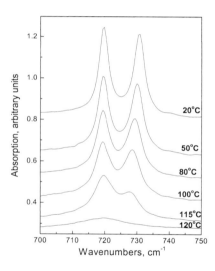

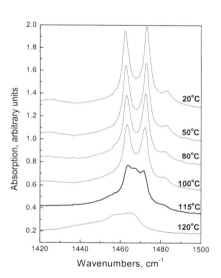

Figure 10. Temperature dependence of methylene rocking mode doublets in 1,44 – Tetratetracosanediol.

Figure 11. Temperature dependence of methylene bending mode doublets in 1,44 – Tetratetracosanediol.

It is known that the temperature-induced phase transition from the orthorhombic to hexagonal packing is accompanied by an increase in the volume of the initial ortho-subcell resulting from thermal expansion of a crystal. X-ray studies[40] have shown that, on thermal expansion, parameter a of the subcell experiences larger changes than parameter b.

It has been shown[39] that high-frequency components of doublets are polarized along the a parameter, while low-frequency components are polarized along the b parameter.

Therefore, it can be expected that the high-frequency components of doublets will be most sensitive to external factors, which was indeed observed in a number of studies of n-paraffins and polyethylene[41, 42].

As follows from Figures 8-11, the high-frequency component in D-22 and D-44 is also most sensitive to temperature. Intensities of bands at 730 cm^{-1} and 1473 cm^{-1} gradually decrease, and eventually these bands vanish. The maxima of the high-frequency components shift toward lower frequencies by several cm^{-1}. The intensities of low-frequency components also gradually decrease, and their maxima shift toward higher frequencies.

Since both doublets consist of overlapping bands, correct quantitative estimates of changes in intensities, half widths of bands and frequency shifts of maxima can be obtained by using computer fitting of the doublets. These investigations are in progress, and detailed results are to be published.

It can be seen from Figures 8-9 that, at the temperature of solid-state phase transition $T_{s-s} T_{max1} =$ 94 °C, splitting of both doublets for D-22 disappears, and, instead, single broad bands with maxima at 720.9 cm^{-1} and 1467.3 cm^{-1} arise. As the melting temperature $T_{max2} = 103$ °C is approached, a weak decrease in the frequencies of maxima (720.4 cm^{-1} and 1467.0 cm^{-1}) is observed. This tendency is preserved at $T > T_{max2}$.

It should be noted that the band at 1467 cm^{-1} of D-22 is asymmetric on the low-frequency side already when it arises at $T = T_{s-s} = 94$ °C (Figure 9). This indicates that this band consists of several overlapping bands. As temperature is increased, especially to $T > T_m$, asymmetry continues to increase. Detailed analysis of this complicated profile is currently being carried out, and the results will be published elsewhere.

For D-44 in the heating cycle, DSC (Figure 3) has revealed only the order-disorder transition (melting) at true $T_m = 117$ °C. FTIR spectroscopic data (Figure 10) show that up to $T \cong 118$ °C the band faintly resembling a doublet is observed at 719.6 cm^{-1}/727 cm^{-1}. But at $T = 120$ °C (true $T_m + 3$ K) it becomes a broad band with a maximum at about $v = 719.7$ cm^{-1}. At higher temperatures (130 °C), an extremely wide band with a maximum at $v = 719.5$ cm^{-1} is seen.

Of interest is the behavior of the doublet at 1463.7/1471.1 cm^{-1} in D-44 (Figure 11). At $T = 115$°C, i.e., below T_m, a third peak at $v = 1466.6$ cm^{-1} appears, in addition to the doublet. It has nearly the same intensity and corresponds to the CH$_2$ bending mode band of an isolated trans-zigzag.

This result is rather important because it indicates that the regions with the low-temperature orthorhombic subcells and regions with the hexagonal subcells co-exist in D-44 below T_m. It is likely that phase transition does not occur simultaneously throughout the entire lamellar volume; a new phase arises in the form of separate domains in the volume of the old phase.

As temperature is further increased to $T = 120$ °C, the doublet disappears, and only a single strongly anisotropic band (on the low-frequency side) is observed. It requires computer fitting, as the band for D-22.

The frequencies at the maxima of single bands (at $v \cong 720$ cm^{-1} and $v \cong 1467$ cm^{-1}) are typical of hexagonal molecular packing of n-paraffins[42]. As known[36], a considerable weakening of the intermolecular interaction occurs at the ortho-hexagonal phase transition due to an increase in the subcell volume. As temperature increases, molecules get sufficient energy to oscillate, twist or rotate about their two-fold screw axes, to such an extent that their cross sections become statistically circular. The basal ab-plane expands so as to make the packing of the chains equal to closest packing of circular rods, i.e., hexagonal subcell.

Of particular interest is analysis of the behavior of absorption bands associated with end hydroxyl groups. In this paper, we analyze only the broad band in the region from 3000 cm^{-1} to 3600 cm^{-1}, which is due to absorption of the O-H stretching vibration of hydroxyl groups. The behavior of other bands and, in particular, the band at $v = 1067$ cm^{-1} (C-O stretching mode) will be analyzed in a separate paper.

It has been shown[43] that for single (monomer) hydroxyl groups, the O-H stretching mode manifests itself in the form of a very narrow band at $v_0 \cong 3650$ cm^{-1}. Single-bridge dimers (1) or double-bridge dimers (2) can arise if hydrogen bonds of the type

are formed. In this case, the (O-H) bond participating in formation of the hydrogen bond becomes weaker, which gives rise to absorption bands at $v_1 = 3623$ cm^{-1}, $v_2 = 3496$ cm^{-1}, and v_3 between 3623 cm^{-1} and 3378 cm^{-1}.

When polymer chains consisting of hydrogen bonds of the type

are formed, the most appreciable weakening of the (O-H) bond in the polymer sequence occurs: the frequency decreases to v_4, which can differ from v_0 by several hundred cm^{-1}[38]. The change Δv of vibration frequencies of (O-H) bonds is proportional to the change in the energy of hydrogen bonds when the distance between diol molecules varies.

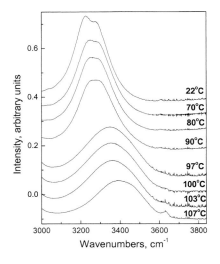

Figure 12. FTIR spectra of the O-H stretching mode of hydroxyl end groups in 1,22 – Docosanediol.

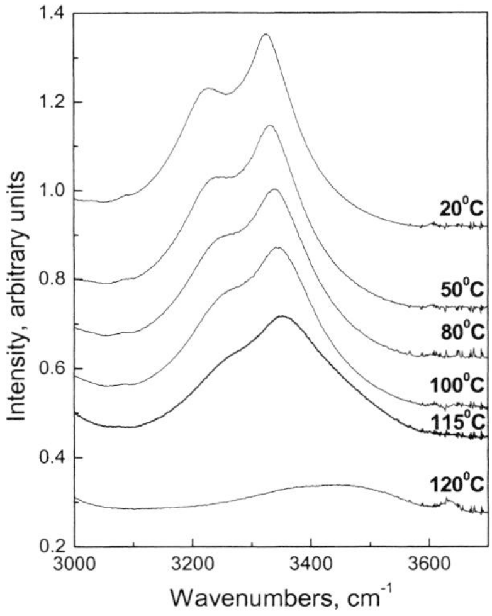

Figure 13. FTIR spectra of the O-H stretching mode of hydroxyl end groups in 1,44 – Tetratetracosanediol at different temperatures.

At room temperature, the absorption spectra of D-22 and D-44 shown in Figure 12 and Figure 13 have two pronounced broad overlapping bands with frequencies of the maxima

$$D\text{-}22: \nu \cong 3225 \text{ cm}^{-1} \text{ and } \nu \cong 3275 \text{ cm}^{-1}$$

$$D\text{-}44: \nu \cong 3225 \text{ cm}^{-1} \text{ and } \nu \cong 3325 \text{ cm}^{-1}$$

in the region from 3000 cm^{-1} to 3600 cm^{-1}. These frequencies differ considerably from $\nu \cong 3650$ cm^{-1} for a monomer and, no doubt, are the evidence of formation of polymer sequences consisting of hydrogen bonds on lamellar surfaces. We suppose that the presence of two maxima with different frequencies is due to formation of polymers sequences in different crystallographic directions of the orthorhombic subcell.

As temperature is increased, the spectrum of D-22 (Figure 12) undergoes appreciable changes well below the phase transition temperature $T_{s-s} = 94$ °C: two bands gradually merge into one broad band and its maximum shifts toward higher frequencies. These effects point to weakening and averaging of the strength of hydrogen bonds, which is, evidently, caused by thermal expansion of the ortho-subcell. This suggests that the frequency $v = 3225$ cm^{-1} in D-22 is attributable to polymer chains consisting of hydrogen bonds formed along the a parameter of the orthorhombic subcell.

Changes at $T > T_{s-s}$ are even more appreciable: at $T = 97$ °C there is only a single very wide band with the maximum at about $v \cong 3350$ cm^{-1} (D-22 has hexagonal subcells). On further heating ($T = 107$ °C $> T_m$), the band half width becomes even wider, and the maximum shifts to $v \cong 3400$ cm^{-1}. However, this frequency still differs significantly from $v \cong 3650$ cm^{-1} for a monomer.

Thus, in spite of transition of D-22 into the melt, and, hence, a higher molecular mobility, end hydroxyl groups of the molecules turn out to be linked by hydrogen bonds. Note that a pronounced single band at $v \cong 3630$ cm^{-1} appears in the spectrum of the melt. According to[43], vibrations with frequencies $v \cong 3350$ cm^{-1} ÷ 3500 cm^{-1} and $v \cong 3630$ cm^{-1} indicate that the melt can contain agglomerates of molecules of diols linked by single-bridge or double-bridge dimers.

The spectra of D-44 (Figure 13) at room temperature also exhibit two rather wide overlapping bands with $v \cong 3225$ cm^{-1} and $v \cong 3325$ cm^{-1}, which indicate, like for D-22, that polymer chains consisting of hydrogen bonds are formed. In contrast to D-22, heating of D-44 does not give rise to a phase T_{s-s} transition of the ortho-hexagonal subcell type, and, therefore, two wide absorption bands are observed up to $T \cong T_m$. This is an additional proof of a higher stability of orthorhombic crystals in D-44, the methylene sequences in which are twice as long as those in D-22. At $T > T_m$ (120 °C and 130 °C), only a single extremely wide band with the maximum in the vicinity of $v \cong 3430$ cm^{-1} is detected. At these temperatures the single band at $v \cong 3635$ cm^{-1} becomes more and more pronounced (like in D-22). Hence, similar to D-22, single-bridge or double-bridge dimers are formed between neighboring diol molecules in the melt of D-44 (with participation of hydrogen bonds).

Conclusions

Similarly to other long-chain molecular crystals, α,ω-alkanediols $(CH_2)_{22}OH$ and $(CH_2)_{44}(OH)_2$ exhibit a well developed morphology, characterized by the presence of large stacks of superposed crystalline lamellae with the thickness proportional to the molecule length.

Splitting of the IR absorption bands into doublets at 720-730 cm^{-1} and 1463-1473 cm^{-1} indicates that crystalline subcells in the lamellae of both diols are of the orthorhombic type, with the parameters typical of hydrocarbons.

Analysis of the DSC data obtained in experiments where the heating rate was varied from $V = 1.25$ K·min^{-1} to $V = 10$ K·min^{-1} allowed estimation of the true solid-state transition temperatures for $(CH_2)_{22}(OH)_2$ ($T_{s-s} = 367.1$ K) and melting temperatures (order-disorder transition) for $(CH_2)_{22}(OH)_2$ ($T_m = 376.3$ K) and $(CH_2)_{44}(OH)_2$ ($T_m = 390.0$ K). It has been shown that hysteresis of T_m in the heating-cooling cycles observed in a number of studies was caused by experimental inaccuracies.

Upon heating of $(CH_2)_{22}(OH)_2$ in the range $T_{s-s} < T < T_m$, splitting of the IR bands at 720-730 cm^{-1} and 1463-1473 cm^{-1} disappears and, instead, only single bands are observed. This indicates that transformation of the orthorhombic subcell into a hexagonal one occurs at the phase transition temperature (T_{s-s}). This type of molecular chain packing remains the same up to the melting temperature T_m.

In a $(CH_2)_{44}(OH)_2$ diol, the ortho-hexagonal subcell transition occurs only at the melting temperature (390.0 K).

Analysis of the wide IR absorption band in the region from 3000 cm^{-1} to 3600 cm^{-1} has led to the conclusion that end hydroxyl groups of diol molecules form on the surfaces of lamellar crystals long (polymer) regular sequences consisting of intermolecular hydrogen bonds. These sequences are arranged in different crystallographic directions.

Linking of the ends of diol molecules by hydrogen bonds becomes weaker on heating, but it is preserved even in the melt. In the melt, single-bridge or double-bridge dimers are formed between neighboring diol molecules (with participation of hydrogen bonds).

Acknowledgements

The authors would like to acknowledge financial support of the Russian Foundation for Basic Research (grant No. 01-03-32773).

[1]   J. D. Hoffmann, B. F. Decker, *J.Phys.Chem.* **1953**, *57*, 520.
[2]   M. G. Broadhurst, *J. Res. Nat. Bur. Stands, A, Phys. and Chem.* **1962**, *66A*, 241.
[3]   P. K. Sullivan, *J. Res. Nat. Bur. Stands, A, Phys. and Chem.* **1974**, *78A*, 129.
[4]   V. Daniel, *Phil.Mag, Suppl.* **1953**, 450.
[5]   P. J. Flory, A. Vrij, *J. Am. Chem. Soc.* **1963**, *85*, 3548.
[6]   M. F. Mina, T. Asano, T. Takahashi, I. Hatta, K. Ito, Y. Amemiya, *Jpn. J. Appl. Phys.* **1997**, *36*, pt.1, 5616.
[7]   J. Techoe, D. C. Bassett, *Polymer* **2000**, *41*, 1953.
[8]   G. Strobl, B. Ewen, E. W. Fischer, W. Piesczek, *J. Chem. Phys.* **1974**, *61*, 5257; *J. Chem. Phys.* **1974**, *61*, 5265. [9] G. R. Strobl, *J. Polym. Sci., Polym. Symp.* **1977**, *59*, 121.
[10]   H. L.Casal, D. G.Cameron, H. H.Mantsch, *Can. J. Chem.* **1983**, *61*, 1736.
[11]   C. Chang, S. Krimm, *J.Polym. Sci., Polym. Phys. Ed.* **1979**, *17*, 2163.
[12]   G. Ungar, N. Masic, *J. Phys. Chem.* **1985**, *89*, 1036.
[13]   G. Ungar, *J. Phys. Chem.* **1983**, *87*, 689.
[14]   A. S. Vaughan, G. Ungar, D. C. Bassett, A. Keller, *Polymer* **1985**, *26*, 726.
[15]   T. Asano, M. F. Mina, I. Hatta , *J. Phys. Soc. Japan* **1996**, *65*, 1699.
[16]   H. Kabayashi, N. Nakamura, *Cryst. Res. Technol.* **1995**, *30*, 495.
[17]   N. Nakamura, S. Setodoi, *Acta Cryst.* **1997**, *C53*, 1883.
[18]   N. Nakamura, T. Yamamoto, *Acta Cryst.* **1994**, *C50*, 946.
[19]   Y. Ogawa, N. Nakamura, *Bull. Chem. Soc. Jpn.* **1999**, *72*, 943.
[20]   R. Popovitz-Biro, J. Majewski, L. Margulis, S. Cohen, L. Leiserowitz, M. Lahav, *J. Phys. Chem.* **1994**, *98*, 4970.
[21]   D. M. Small, "The Physical Chemistry of Lipids", Plenum, N.Y.-London 1986.
[22]   C. Le Fevere de Ten Hove, A. Jonas, J. Penelle, *Proc .Am. Chem. Soc. Div. Polym. Mater. Sci. Eng.* **1997**, *76*, 158.
[23]   E. E. Rusanova, Y. L. Sebyakin, L. V. Volkova, R. P. Evstigneeva, *J. Org. Chem. USSR*, **1984**, *20*, 248.
[24]   B. G. Ranby, F. F. Morehead, N. M. Walter, *J. Polym. Sci.* **1960**, *44*, 349.
[25]   A. Keller, *Phil. Mag.* **1961**, *6*, 329.
[26]   H. Zocher, R. D. Machado, *Acta Cryst.* **1959**, *12*, 122.
[27]   I. M. Dawson, V. Vand, *Proc. Royal Soc., London, Ser.A.* **1951**, *206A*, 555.
[28]   S. Amelinckx, *Acta Cryst.* **1956**, *9*, 217.
[29]   I. M. Dawson, *Proc. Royal Soc. London,* **1952**, *A214*, 72.
[30]   L. D. Landau, E. M. Liphshitz, *"Statistical Physics"*, Nauka, Moscow 1976, Ch. XIV.

338

[31]   V. A. Bershtein, V. M. Egorov, *"Differential Scanning Calorimetry of Polymers: Physics, Chemistry, Analysis, Technology"*, Ellis Horwood, N.Y., 1994.
[32]   J. Sestak ,*"Thermophysical Properties of Solids"*, Academia, Prague 1984.
[33]   M. Fisher, *"Nature of critical state"*, Mir, Moscow 1973.
[34]   L. A. Nikolaev, V. A. Tulupov, *"Physical Chemistry"*, Khimiya, Moscow 1967.
[35]   C. W. Bunn, *Trans. Far. Soc.* **1939**, *35*, 482.
[36]   A. Mueller, *Proc. Roy. Soc.* **1932**, *138*, 514.
[37]   S. Krimm, *Adv. Polym. Sci.* **1960**, *2*, 51.
[38]   K. Nakamishi, *"Infrared Absorption Spectroscopy"*, Holden-Day, San Francisco 1962, Ch.II.
[39]   R. G. Snyder, *J. Mol. Spect.* **1961**, *7*, 116.
[40]   G. T. Davis, R. K. Eby, G. M. Martin, *J. Appl. Phys.* **1968**, *39*, 4973.
[41]   H. L. Casal, H. H. Mantsch, D. G. Cameron, R. G. Snyder, *J. Chem. Phys.* **1982**, *77*, 2826.
[42]   J. R. Nielsen, C. E. Hathaway, *J. Mol. Spectr.* **1963***, 10*, 366.
[43]   F. A.Smith, E. C. Creitz, *J. Res. Nat. Bur. Stands.* **1951**, *46*, 145.

Synthesis, Characterization and Properties of Novel Poly(Ester-Amide-Urethane)s

Shahram Mehdipour-Ataei, *Parvin Einollahy*

Iran Polymer and Petrochemical Institute, P.O.Box 14965/115, Tehran, Iran

E-mail: s.mehdipour@proxy.ipi.ac.ir

Summary: A diacid (TOBA) containing an ester group was synthesized by reaction of terephthaloyl chloride with 4-hydroxybenzoic acid. Reaction of the obtained diacid with thionyl chloride resulted in preparation of the related diacid chloride (TOBC). Nucleophilic substitution reaction of 4-aminophenol and also 5-amino-1-naphthol with the prepared diacid chloride afforded two aromatic diols containing ester and amide groups, respectively. Aromatic and semi-aromatic poly(ester-amide-urethane)s were prepared via addition polymerization of different diisocyanates with novel diols. The prepared polyurethanes showed improved thermal stability.

Keywords: addition polymerization; copolymerization; polyurethanes; synthesis; thermal properties

Introduction

Polyurethanes comprise a class of versatile materials that have excellent abrasion resistance and the properties of both rubber and plastics. They are becoming more important as engineering materials.[1] In spite of their extended applications, the main drawback of polyurethanes is their poor heat resistance. Their acceptable mechanical properties disappear at about 80-90°C and thermal degradation takes place above 200°C. [2]

Attempts to improve the thermal stability of polyurethanes have been made over a long period. One important method by which to improve the thermal stability of polyurethanes is chemical modification of their structure by copolymerization with highly thermally-stable polymers. This can be achieved by incorporation of an amide and ester moiety into the polyurethane chain. Polyamides and polyesters are important classes of high performance polymers that have remarkable high temperature resistance and superior physical and mechanical properties. [3-5]

Pursuing our continued interest in synthesizing thermally stable polymers, [6-15] here we describe the synthesis of novel diols with preformed ester and amide structures. Addition polymerization of diols with six different diisocyanates led to the preparation of twelve different aromatic and semi-aromatic poly(ester-amide-urethane)s. The presence of urethane, ester, and amide linkages in the polymer chain may alter the properties of polymers to an even greater extent. Hence it was thought interesting to synthesize and characterize poly(ester-amide-urethane)s.

Experimental

Materials

All the required chemicals were purchased either from Merck or Aldrich Chemical Co. Terephthaloyl chloride, p-phenylene diisocyanate (PPDI), naphthalene diisocyanate (NDI), and cyclohexane diisocyanate (CHDI) were purified by sublimation. Diphenylmethane diisocyanate (MDI), toluene diisocyanate (TDI), and dicyclohexylmethane diisocyanate (H_{12}MDI) were purified by distillation. NMP, DMAc, and toluene were purified by vacuum distillation over CaH_2.

Analytical Instruments

Infrared measurements were performed on a Bruker IFS-48 FT-IR spectrometer. The ^{1}H-NMR spectra were recorded in dimethyl sulfoxide (DMSO-d_6) solution, using a Bruker Avance DPX 500-MHz instrument. Elemental analyses were performed by a CHN-O-Rapid Heraeus elemental analyzer. Differential scanning calorimetry (DSC) and thermogravimetric analysis (TGA) were recorded on a Stanton Redcraft STA-780. Differential thermogravimetric (DTG) traces were recorded on a Polymer Lab TGA-1500. The dynamic mechanical measurements were performed on a Polymer Laboratories Dynamic Mechanical Thermal Analyzer (Model MK-II) over a temperature range of −100 to 200 °C at 1 Hz and a heating rate of 5 °C /min. The value of tan δ and the storage modulus versus temperature were recorded for each sample. Inherent viscosities were measured using an Ubbelohde viscometer.

Monomer Synthesis

Synthesis of terephthaloyl bis (4-oxybenzoic)acid (TOBA):

4-Hydroxybenzoic acid (23.14g) was dissolved in 350 ml of a 0.1 M aqueous solution of sodium hydroxide. After cooling the solution to 10°C, 11.58g of terephthaloyl chloride dissolved in 115 ml of tetrachloroethane and was added drop-wise to the mixture with stirring. The reaction mixture was stirred at room temperature for 6 h. The mixture was soaked in 430 ml of 3M HCl for 14h.

The product was obtained by filtration and washed with water, ether, and ethanol. Then it was vacuum dried at 60°C overnight (yield 85%).

[IR (KBr): v 3250 (O-H); 1748 (-COOR); 1688 (-COOH); 1425 (C=C); and about 1285 cm^{-1} (C-O); ^{1}H-NMR (ppm) (DMSO-d$_6$): δ 13.11 s (2H, COOH); 8.04-8.27 m (8H, aromatic); 7.45-7.80m (4H, aromatic)].

Synthesis of terephthaloyl dioxydibenzoylchloride (TOBC):

TOBA (7.0 g) was refluxed with thionyl chloride (87.5ml) and two drops of dry DMF for 4 h. Excess thionyl chloride was removed from the mixture by distillation. The crude product was recrystallized from ether and vacuum dried at 50°C. (Yield 89%). [IR (KBr): v 1780 (-COCl); 1745 (-COOR); 1600 (C=C); 1408 (C=C); 1265 cm^{-1} (C-O).

Synthesis of diol (AP-diol):

4-Aminophenol (0.048 mole) was dissolved in 60 ml of dry NMP. The mixture was cooled to 0°C and 5 ml of propylene oxide was added. TOBC (0.0153 mole) was added to the mixture and stirred at 0°C for 0.5 h. The mixture was further stirred for 4 h at room temperature. The product was obtained by pouring the flask content into 300 ml of H$_2$O. It was then filtered and washed with methanol. (Yield 88%). [IR (KBr): v 3320-3345 (NH, OH); 1734 (-COOR), 1643 (-CONH), 1250 cm^{-1}(C-O); ^{1}H-NMR(ppm) (DMSO-d$_6$): δ 10.10 s (2H, NH); 9.22 s (2H,OH); 8.06-8.29 m (8H, aromatic); 7.49-7.56 m (8H, aromatic); 6.76-6.78 m (4H, aromatic).

The same procedure was performed to prepare AN-diol. In this case, 5-amino-1-naphthol was used instead of 4-aminophenol. (Yield 86%). [IR (KBr): v 3198-3250 (NH, OH); 1736(-COOR), 1641 (-CONH), 1263 cm^{-1} (C-O); ^{1}H-NMR (ppm) (DMSO-d$_6$): δ 10.16 s (2H, NH); 9.40 s (2H, OH); 8.07-8.30 m (8H, aromatic); 7.20-7.57 m (12H, aromatic); 6.58-6.62 m (4H, aromatic).

Polyurethane Synthesis

The synthesis of a polyurethane was typically carried out as follows: Diisocyanate (0.0012 mole) was dissolved in 5 ml of dry NMP under a nitrogen atmosphere. Two drops of dibutyltin dilaurate was added to the mixture and heated to 110°C . Then 0.0012 mole of the prepared diol (AP or AN diol) was dissolved in 5 ml of dry NMP, added to the mixture and heated for 4 h. The mixture was cooled, poured into 200 ml of H$_2$O and filtered. Then the polymer was purified using DMF and H$_2$O as a solvent-non solvent system. The obtained polymer was vacuum dried at 70°C for 5 h. Yield of the reactions was about 73-95%.

Results and Discussion

Improving the thermal stability of polyurethane was the main aim of this study. This was achieved by the following structural modifications: (a) introduction of thermally stable amide and ester groups into the polymer backbone; (b) phenylation of the backbone; (c) retention of regularity and symmetry of units.

Thus, preparation of two novel, fully aromatic diols with preformed ester and amide groups was considered. In this way, by condensation reaction of 4-hydroxybenzoic acid with terephthaloyl chloride, according to the Schotten-Baumann reaction, a diacid (TOBA) was prepared (Scheme 1).

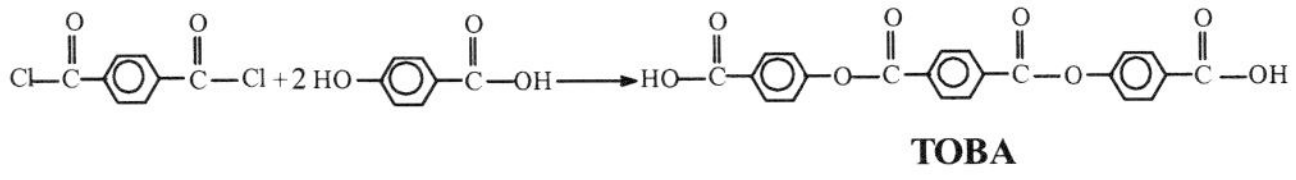

Scheme 1. Synthesis of TOBA

Conversion of the diacid to the corresponding diacid chloride (TOBC) was achieved by SOCl$_2$ in the presence of catalytical amount of dry DMF (Scheme 2).

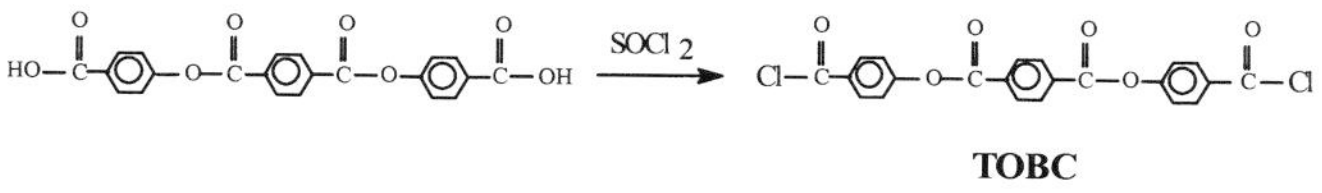

TOBC

Scheme 2. Synthesis of TOBC

Two novel aromatic diols containing built in ester and amide units were obtained by the nucleophilic substitution reaction of TOBC with 4-aminophenol and also 5-amino-1-naphthol (Scheme 3).

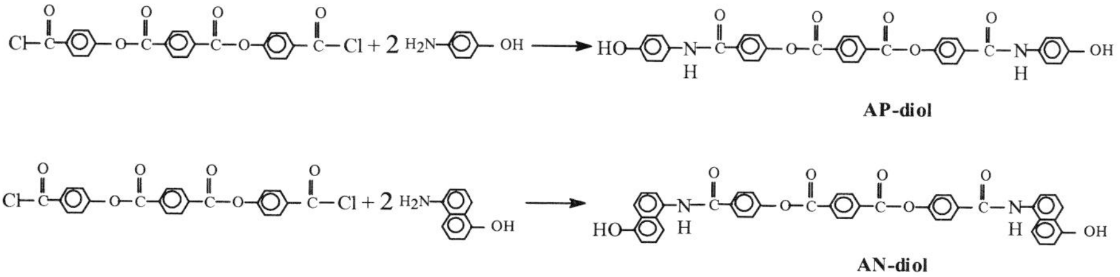

AP-diol

AN-diol

Scheme 3. Preparation of diols with preformed groups

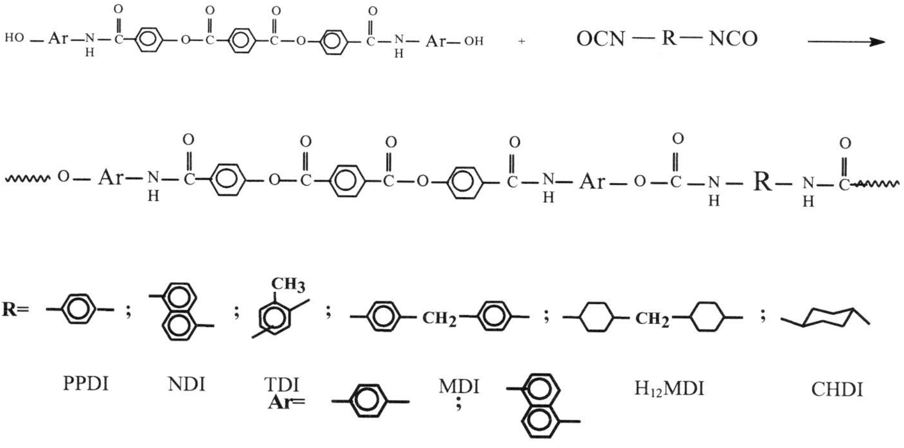

R= PPDI ; NDI ; TDI ; MDI ; H$_{12}$MDI ; CHDI

Ar=

Scheme 4. Preparation of poly (ester-amide-urethane)s

Addition polymerization of the prepared diols with aromatic and aliphatic diisocyanates afforded 12 different poly(ester-amide-urethane)s. (Scheme 4).

Structures of the prepared polyurethanes and yields of the reaction are tabulated in Table 1.

Table 1. Structures of the poly (ester-amide-urethane)s

Reactants	Structures of the polymers	Yield (%)
1. Diol- AP+ PPDI		86
2. Diol- AP+ CHDI		90
3. Diol- AP+ NDI		76
4. Diol- AP+ TDI		97
5. Diol- AP+ MDI		90
6. Diol- AP+ H_{12}MDI		94
7. Diol- AN+ PPDI		80
8. Diol- AN+ CHDI		84
9. Diol- AN+ NDI		73
10. Diol- AN+ TDI		88

| 11. Diol- AN+ MDI | | 90 |
| 12. Diol- AN+ H$_{12}$ MDI | | 95 |

The polyurethanes were characterized by spectroscopic methods and the results are tabulated in Table 2.

Table 2.   Spectroscopic analysis of the polyurethanes

Polyurethane	IR(KBr)cm^{-1}	^{1}H-NMR(DMSO-d$_6$) δ
1)AP+PPDI	3323, 1734, 1643, 1610, 1541, 1252, 866	10.47 (amide,2H), 10.30 (urethane,2H), 6.71-8.32 (aromatic,24H)
2) AP+ CHDI	3327,2928, 1730, 1643, 1539, 1514, 1252, 827	10.18 (amide,2H), 9.30 (urethane,2H), 6.76-8.29 (aromatic,20H) 3.28-3.30 (CH,2H), 2.18-2.20 (CH$_2$,8H)
3)AP + NDI	3325, 1736, 1644, 1539, 1516, 1254, 834	10.45 (amide,2H), 10.31 (urethane,2H), 6.70-8.51 (aromatic,26H)
4) AP + TDI	3321, 2935,1735, 1646, 1537, 1323, 1252, 826	10.46 (amide,2H), 10.30 (urethane,2H), 6.72-8.39 (aromatic,23H) 1.78 (CH$_3$,3H)
5) AP + MDI	3329,2920, 1738, 1643, 1609, 1544, 1250, 864	10.16 (amide,2H), 9.29 (urethane,2H), 6.75-8.52 (aromatic,28H) 3.90 (CH$_2$,2H)
6) AP+H$_{12}$MDI	3325,2922, 1732, 1647, 1530, 1517, 1253, 866	10.21 (amide,2H), 9.27 (urethane,2H), 6.72-8.26 (aromatic,20H) 3.17 (aliphatic,2H), 2.09-2.13 (aliphatic,18H), 1.30(CH$_2$,2H)
7) AN + PPDI	3302, 1736, 1641, 1597, 1514, 1406,1269, 826	10.46 (amide,2H), 10.27 (urethane,2H), 6.70-8.35 (aromatic,28H)
8) AN+CHDI	3319,2928, 1745, 1642, 1597, 1531, 1268, 861	10.19 (amide,2H), 9.29 (urethane,2H), 6.70-8.51 (aromatic,24H) 3.29-3.32 (CH,2H), 2.18-2.20 (CH$_2$,8H)
9) AN+ NDI	3320, 1731, 1644, 1538,	10.47 (amide,2H), 10.30 (urethane,2H), 6.71-8.53 (aromatic,30H)

	1510, 1252, 836	
10) AN + TDI	3306,2972, 1742, 1643, 1597, 1531, 1267, 869	10.47 (amide,2H), 10.32 (urethane,2H), 6.71-8.61 (aromatic,27H) 1.78 (CH_3,3H)
11) AN+ MDI	3300,2968, 1736, 1648, 1611, 1511, 1257, 856	10.15 (amide,2H), 9.30 (urethane,2H), 6.73-8.53 (aromatic,32H) 3.87 (CH_2,2H)
12) AN+H_{12}MDI	3323, 2922,1732, 1642, 1539, 1513, 1250, 865	10.33 (amide,2H), 9.97 (urethane,2H), 6.72-8.26 (aromatic,24H) 3.20 (aliphatic,2H), 2.11-2.17 (aliphatic,18H), 1.32(CH_2,2H)

The results of elemental analysis were in good agreement with the calculated amounts and, therefore, confirmed the structures. (Table 3).

Table 3.   Elemental analysis of the polyurethanes

	Polyurethane	Calculated			Found		
		C	H	N	C	H	N
1.	AP-PPDI: $C_{42} H_{28} N_4 O_{10}$	67.38	3.74	7.49	67.17	3.65	7.62
2.	AP-CHDI: $C_{42} H_{34} N_4 O_{10}$	66.84	4.51	7.43	66.75	4.23	7.56
3.	AP-NDI: $C_{46} H_{30} N_4 O_{10}$	69.17	3.76	7.02	69.10	3.66	7.08
4.	AP- TDI: $C_{43} H_{30} N_4 O_{10}$	63.54	3.69	6.90	63.59	3.51	6.98
5.	AP-MDI: $C_{49} H_{34} N_4 O_{10}$	70.17	4.06	6.68	70.15	4.10	6.53
6.	AP-H_{12}MDI: $C_{49} H_{46} N_4 O_{10}$	69.18	4.00	6.59	69.01	3.95	6.72
7.	AN-PPDI: $C_{50} H_{32} N_4 O_{10}$	70.75	3.77	6.60	70.60	3.68	6.73
8.	AN-CHDI: $C_{50} H_{38} N_4 O_{10}$	70.26	4.45	6.56	70.30	4.51	6.67
9.	AN-NDI: $C_{54} H_{34} N_4 O_{10}$	72.16	3.79	6.24	72.09	3.85	6.20
10	AN-TDI: $C_{51} H_{34} N_4 O_{10}$	70.99	3.94	6.50	70.81	3.85	6.63

| 11 | AN-MDI: $C_{57} H_{38} N_4 O_{10}$ | 72.92 | 4.05 | 5.97 | 73.01 | 4.09 | 6.01 |
| 12 | AN-H_{12}MDI: $C_{57} H_{50} N_4 O_{10}$ | 72.00 | 5.26 | 5.89 | 71.92 | 5.20 | 5.98 |

The inherent viscosity of the polymers measured at a concentration of 0.5 g/dl in DMAc at 30°C (Table 4) was about 0.35-0.42 dl/g, indicating acceptable molecular weights.

Solubility of the prepared polyurethanes in dipolar aprotic solvents was about 0.5-1.3 g/dl.

Thermal properties and thermal behavior of the polymers was studied using DSC, TGA, and DMTA techniques. The T_g of the polymers was in the range of 155-179°C, the 10% weight loss was in the range of 278-299°C, and the char yields of the polymers at 550°C were about 24-48% (Table 4).

According to the results, the thermal stabilities of the prepared polyurethanes are higher than the conventional polyurethanes. This can be mainly attributed to the incorporation of ester and amide groups into the polymer backbone.

Thus, polyurethanes derived from AP-diol are more heat resistant and less soluble than the AN-diol derived polyurethanes. This is a result of symmetry and close packing of the aminophenol unit in comparison with the bulky aminonaphthol unit.

On the other hand, polyurethanes derived from fully aromatic diisocyanates are more heat resistant and less soluble than the polyurethanes derived from aliphtic diisocyanates.

Table 4.   Thermal analysis and viscosity of the polymers

Polyurethane	T_g (°C)	T_0 (°C)	T_{10} (°C)	T_{max} (°C)	Char Yield (at 550°C)	Inherent Viscosity (dL/g)
1)AP+PPDI	179	190	299	327	33	0.42
2) AP+ CHDI	170	180	287	325	28	0.41
3)AP + NDI	176	188	290	309	39	0.37
4) AP + TDI	171	179	288	306	29	0.38
5) AP + MDI	172	185	293	322	34	0.40
6) AP+H_{12}MDI	168	177	284	315	24	0.40
7) AN + PPDI	177	189	297	322	48	0.38
8) AN+CHDI	158	175	282	323	37	0.35
9) AN+ NDI	172	185	289	310	38	0.35
10) AN + TDI	166	177	284	300	26	0.36
11) AN+ MDI	169	183	287	317	33	0.37
12) AN+H_{12}MDI	155	169	278	303	25	0.39

In general, more aromatic, more symmetric and less bulky polyurethanes have high thermal stability and low solubility. Also, more aliphatic, more bulky and less symmetric polyurethanes have high solubility and low thermal stability.

Conclusions

To improve the thermal stability of polyurethanes, introduction of thermally stable units into the polymer backbone was considered. In this way, two aromatic diols with preformed ester and amide groups were prepared. Polyaddition reactions of the two diols with different diisocyanates afforded poly(ester-amide-urethane)s. The polymers were characterized and their physical properties were examined. All the polyurethanes showed improved thermal stability in comparison with conventional polyurethanes.

[1]   G. Ortel, *"Polyurethane Handbook"*, Hanser Publishers, New York, 1985, Chapter 2.
[2]   B. Masiulanis, R. Zielinski, *J. Appl. Polym. Sci.* **1985**, *30*, 2731.
[3]   P. W. Morgan, *"Condensation Polymers by Interfacial and Solution Methods"*, Polym. Rev. Series X, John Wiley & Sons Inc. New York, 1965.
[4]   F. Higashi, S. Ogata, Y. Aoki, *J .Polym. Sci. Part A: Polym. Chem.* **1982**, *20*, 2081.
[5]   G.C. Wu, H. Tanaka, K. Sanui, N. Ogata, *Polym. J.* **1982**, *14* ,797.
[6]   M. Barikani, S. Mehdipour-Ataei, *J. Polym. Sci. Part A: Polym. Chem.* **1999**, *37*, 2245.
[7]   S. Mehdipour-Ataei, M. Barikani, *Iranian Poly. J.* **1999**, *8*, 3.
[8]   M. Barikani, H.Yeganeh, S. Mehdipour-Ataei, *Polym. Inter.* **1999**, *48*, 1264.
[9]   M. Barikani, S. Mehdipour-Ataei, *J. Polym. Sci. Part A.. Polym. Chem.* **2000**, *38*, 1487.
[10]   H. Yeganeh, S. Mehdipour-Ataei, *J. Polym. Sci. Part A: Polym. Chem.* **2000**, *38*, 1528.
[11]   M. Barikani, S. Mehdipour-Ataei, *J. Appl. Polym. Sci.* **2000**, *77*, 1102.
[12]   M. Barikani, S.Mehdipour-Ataei, H.Yeganeh, *J. Polym. Sci. Part A. Polym. Chem.* **2001**, *39*, 514.
[13]   S. Mehdipour-Ataei, *Iranian Poly. J.* **2002**, *11*, 251.
[14]   S. Mehdipour-Ataei, S. Keshavarz, *J. Appl. Polym. Sci.* **2003**, *88*, 2168.
[15]   S. Mehdipour-Ataei, H. Heidari, *Macromol. Symp.* **2003**, *193*, 159.
[16]   J.H. Saundres, K.C. Frisch, *"Polyurethane, Part 1 Chemistry"*, Interscience, New York, 1962, Chapter 6.
[17]   I. Keijj, "Handbook of *Polyurethane Resins"*, The Nikkan Kogyo Shimmbun LTD, Japan, 1987, Chapter 1.
[18]   F.L. Lin, F.S. Chuang, Y.C. Shu, *Polym. Plast. Technol. Eng.*, **1998**, *37*, 71.

Self-Diffusion of PEO-Modified Paclitaxel in Aqueous Solution: Hydrodynamic Properties

Dedicated to Prof. *Jung-Il Jin*, Korea University, Seoul, Korea, on the occasion of his 60th birthday

Michael Hess,[*1] *Manfred Zähres,*[1] *Byung-Wook Jo*[2]

[1] Department of Physical Chemistry, Gerhard-Mercator-University, 47048-Duisburg, Germany
E-mail: hi259he@uni-duisburg.de
[2] Department of Polymer Science and Engineering and Department of New Materials in Biology, Chosun University, 375 Seosuk-dong, Dong-gu, GwangJu, 501-759, South Korea

Summary: The anti-cancer drug Paclitaxel has been hydrophilized by coupling with poly(ethylene oxide) through a self-immolating linker to the polymer. The mobility of the functionalized polymer and of the amphiphilic drug-modified polymer in D_2O was studied by the temperature and concentration dependence of the long-time self-diffusion coefficients of the components. The technique applied was pulse-gradient field NMR spectroscopy. The diffusivity of the molecules can be described in terms of the Einstein-Sutherland and the Stokes-Einstein equations. First results indicate that combining the drug with the polymer rather influences the rigidity parameter and the pair-interaction in solution than the shape and the hydrodynamic radius of the polymer coil. Unexpectedly, micelle-formation, although highly probable, was not observed within the concentration range investigated here.

Keywords: amphiphilic polymer; mobility; pulse-gradient field NMR

Introduction

Paclitaxel is a lipophilic anti-cancer active drug that can be isolated from the bark of the pacific yew tree (Taxus *brevifolia*). Its anti-cancer activity concentrates on interfering with the mitosis through destabilization of the microtubuli. Paclitaxel is proven to interact in such a manner with a number of different cancer types, such as lung, breast, ovary, and some leukaemias.

As a lipophilic molecule Paclitaxel cannot be applied directly to the body by injection or orally; it has to be hydrophilized first. There are different potential reactive sites for this on the molecule, see Fig. 1. One successful way to hydrophilize a lipophilic molecule is to attach

DOI: 10.1002/masy.200451026

it to a hydrophilic polymer chain like poly(ethylene oxide) (PEO). PEO itself is known for its tumour targeting potential in that PEO accumulates preferably in some malign tissues. The preferable sites for the modification, where PEO can be coupled with Paclitaxel, are positions 2' and 7 are positions 2' and 7, see Fig. 1. This has been successfully done, as described in various patents, e.g.[1]. The disadvantage of these modifications was that the stability of the modified drug was too high. The modified drug has to be released from the polymer by an esterase and, with the types of linkage between the polymer and the drug existing to date, this reaction takes several hours. As a consequence, much of the drug was excreted by the body before it could develop its activity.

Jo et al.[2] recently developed a patented self-immolating link between PEO and the drug which provides a hydrolytic stability of the drug-polymer link in the body liquid of only a few minutes, just sufficient to transport the major amount of the drug to the site where its activity is required.

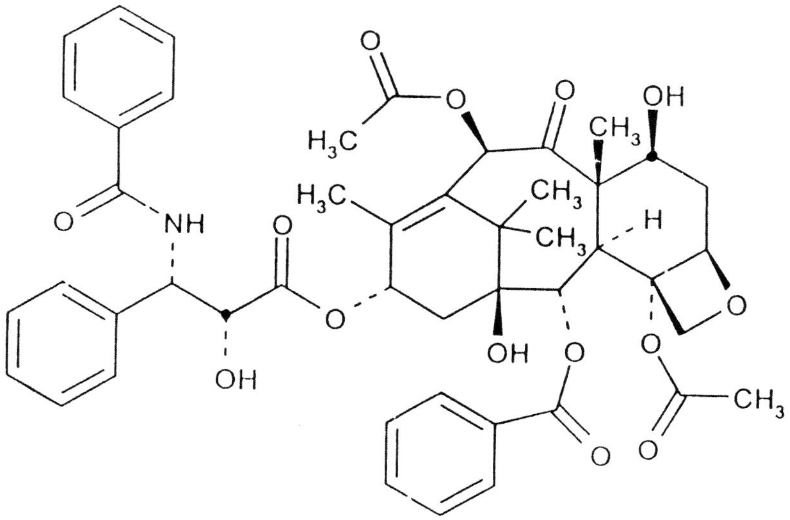

Fig. 1:   The anti-cancer active drug Paclitaxel with its potential docking sites on C-2' and on C-7 for a hydrophilic modification with PEO. (The derivative investigated in this study is the pre-drug with substitution at C-7, termed PP7.)

Further studies of the transport behaviour of the drug in the body liquid require detailed knowledge about the mobility of the molecule, its interaction and phase behaviour under different conditions. Investigations into the surface activity, the search for mesophases to be expected because of the likely amphiphilic character of the molecule, and modifications of the substitution pattern of the drug are presently underway. In later stages of the investigations it will be of particular interest to study possible interactions with proteins present in the body fluid and to consider diffusion in confined environments (as they are provided by cells), therefore knowledge of the "simple" pure systems is important.

In a previous publication[3] we reported first results on diffusion of these systems in D_2O. We now report further results on the investigation of the hydrodynamic properties of the 7'-substituted Paclitaxel (PP7) and succinic acid-modified PEOS (5000 g/mol and 20000 g/mol) as determined from pulse-gradient field NMR/experiments. The monosuccinyl PEO (PEOS) was chosen to compare the behaviour of the polymer coil with the corresponding behaviour of the modified drug. The pulse gradient field technique has been described elsewhere eg. by Callaghan et al.[4], Steijskal and Tanner[5], Otting[6], Stilbs[7] and Kimmich[8]. The Einstein-Sutherland equation (eq. 1) describes the self-diffusion coefficient of Brownian particles:

$$D = \frac{k_b \cdot T}{f} \tag{1}$$

$D \equiv$ self-diffusion coefficient, $f \equiv$ friction factor, $k_b \equiv$ Boltzmann-factor, $T \equiv$ thermodynamic temperature.

In the case of spheres at infinite dilution (that is, for a single particle) eq. 1 becomes the well-known Stokes-Einstein equation:

$$D_0 = \frac{k_b \cdot T}{6 \cdot \pi \cdot \eta_s \cdot R_H^3} \tag{2}$$

$\eta_s \equiv$ the viscosity of the pure solvent, here D_2O, taking into account its temperature dependence; $R_H \equiv$ the hydrodynamic radius of the equivalent sphere.

The volume fraction dependence of the self-diffusion coefficient for a dispersion of colloidal hard spheres can be described by a series expansion in two equivalent ways, see for example Dhont[9] or Russel et al.[10]:

$$D = D_0 (1 - \lambda \cdot \varphi + ...) \tag{3}$$

or through the concentration dependence of the friction factor f:

$$f = f_0(1 + v \cdot \varphi + ...) \tag{4}$$

where f_0 equals $6\pi\eta_s R_H^3$. In the case of polymer chains the volume fraction can best be described by taking the coil volume fraction, which is related to the coil-overlap concentration c^*:

$$\varphi = c/c^* \tag{5}$$

with $c \equiv$ the mass concentration and c^* defined by:

$$c^* = \frac{M}{\dfrac{4}{3} \pi R_H^3 N_A} \tag{6}$$

$M \equiv$ the molar mass; $N_A \equiv$ Avogadro's constant.

At infinite dilution f in eq. 1 is equal to f_0. The coefficient v is sometimes called the rigidity parameter and depends on the hydrodynamic pair interaction between the particles and thus depends on the shape and rigidity of the particles. For rigid spheres $v = 2.00$ in the absence of hydrodynamic interactions, Pyun and Fixman[11], Hanna et al.[12]. Dhont et al.[13] have extended the theory for rod-like particles. In the case of repulsive interactions v becomes smaller than 2, whereas it becomes larger in the case of attractions. This can be understood by realising, that in the case of attractions, two particles of the same component will - on the average - stay longer in proximity with each other than with other molecules, thereby slowing down their diffusion.

Experimental

Paclitaxel, see Fig. 1, was coupled to α-hydro-ω-methyl-polyl(ethylene oxide) (PEO) with a self-immolating succinic acid spacer (abbreviation of the PEO with the linker: PEOS) to obtain the water soluble product PP7, as described elsewhere[14]. A highly uniform PEO of molar mass Mw = 5000 g/mol (PEOS 5000) was used to create the desired water solubility.

The non-uniformity of the molar mass was 1.05 for all polymers, the succinate derivative, PEOS and the hydrophilized drug PP7. There was also a PEOS of molar mass Mw = 20000 g/mol, with the same narrow distribution of the molar mass, available for comparative studies. The modified drug PP7, is however, presently only available attached to the 5000 g/mol polymer.

The experiments were carried out on a Bruker Avance DRX 500 equipped with a water cooled Diff30 probe with z-gradient and a 40 A gradient amplifier. The gradient strength was 0.3 T m^{-1} A^{-1}. The measurements were done in a Bruker 11.74 T standard bore magnet. The pulsed magnetic field gradient magnitude g was varied up to 6 T/m. The gradient was calibrated by comparison with literature data of D$_2$O and was accurate within 1.5%. The sample temperature was determined after calibration with the temperature dependence of the peak shift in methanol. Because of the development of undesired temperature gradients in the probe head which caused convection in the sample, the sample temperature was controlled by the water jacket of the z-gradient coil. The samples were allowed to equilibrate for 30 min prior to the experiments being carried out. The variance in the determined values of the diffusivity was always of the order of magnitude of 2×10^{-13} m^2 s^{-1}. The experiments were carried out in D$_2$O of purity > 99.95 D%.

Results and Discussion

The diffusion time Δ during which the NMR experiments were carried out delivered the so-called "long-term" diffusion coefficient which means that the time scale of the motion is large compared with the relaxation times of the particles[11], that means t > 10^{-6} s. During a diffusion time $\Delta = 350$ ms the diffusion coefficients were determined with an accuracy of about $2 \cdot 10^{-13}$ m^{2}s^{-1}. Fig. 2 shows the concentration and temperature dependence of the diffusion coefficient of PEOS 5000 and PP7. Fig. 3 shows the influence of the higher molar mass on the self-diffusion coefficient of PEOS. Clearly, the molecules interact as indicated by the decreasing self-diffusion coefficient with the polymer concentration. At low concentrations there is an almost perfect linear dependence, whereas at higher concentrations higher coefficients of the expansion have to be considered. The concentration dependence of the self-diffusion coefficient is stronger in case of the drug-modified PEO, indicating attractive interactions between the molecules which lead to an increase of the λ-parameter and thus to an increase in the self-diffusion coefficient. However, in the concentration range investigated here, there is no evidence of critical micell formation. No lyotropic phases were observed in this concentration range.

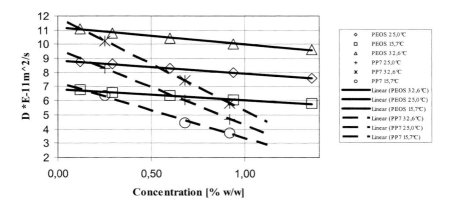

Fig. 2:   Diffusion coefficient D vs. composition with the temperature as parameter. The molar mass of the PEO chain is 5000 g/mol

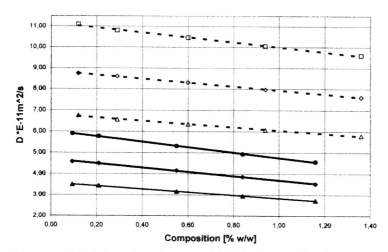

Fig. 3:   Diffusivity and molar mass of PEOS. Composition dependence of the diffusion coefficient with the temperature as parameter. dotted lines: PEOS 5000 g/mol; full lines: PEOS 20000 g/mol, open squares: T = 32,6°C, open lozenges: T = 25°C, open triangles: T = 15°C, filled circles: T = 32,6°C, filled lozenges: T = 25°C, filled triangles: T = 15°C

The temperature dependence of the self-diffusion coefficient is of an Arrhenius-type, with an apparent energy of activation constant within experimental accuracy for all samples in the investigated concentration/temperature range. This indicates no significant changes in the transport mechanism. There is also no significant change in the size and shape of the molecules. Hence, the formation of larger aggregations or micelles could not be proven yet. The hydrodynamic radius R_H at 25°C is calculated with eq. 2 and the results are shown in Table 1.

Table 1: Hydrodynamic radius and series expansion coefficient λ in eq. 3

	R_H/nm	λ
PEOS 5000	2.23	1.2
PEOS 20000	4.13	6.2
PP7	2.06	12

The fact that the single-particle self-diffusion coefficient for dilute solutions, D_0, does not change drastically form PEOS to PP7 indicates that the shape of the molecules does not change significantly and that both types of molecules nearly behave like rigid spheres, as is expected and known for pure PEO[15]. Analysis of the concentration dependence of the self-diffusion coefficient in terms of eq. 3 shows an about 10 times larger expansion coefficient λ compared with the drug-free PEOS 5000, see Fig. 4. Since the shape of the molecules is not significantly altered by the coupling of the drug with the polymer chain as pointed out above, an explanation might be found in a change in the interaction potential[16]. The magnitude of the effect could be explained by effective attractions between the modified drug molecules. The behaviour of PEOS 20000 can be understood as that of a soft sphere.

PEGS-PP7

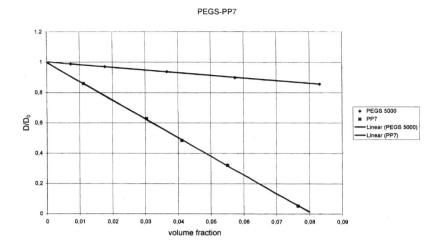

Fig. 4:   Concentration dependence of the diffusion coefficient of PEOS 5000 and PP7 at
25°C in terms of eq. 3

Conclusion

Coupling the hydrophobic Paclitaxel with PEO via a self-immolating linker suggests an amphiphilic macromolecular product consisting of the **hydrophobic** drug as the **"head"** and the **hydrophilic** macromolecular **"tail"**. However, the rather similar D_0-values of PEOS and PP7 (see Fig. 2) indicate that the pre-drug does not behave differently to that of a rigid spheroid like PEO (or PEOS). The apparent hydrodynamic radius of PP7 is only slightly smaller compared with PEOS 5000. The major difference in the diffusion behaviour appears to be caused by the friction factor. This means there are stronger interactions in pre-drug solutions compared with PEOS solutions. Since there were no lyotropic phases or micelles observed in the concentration range investigated here, a probable interpretation is that the polymer chain attached to the pre-drug coils around the drug, resulting in a core-shell-like structure comparable to the folding of many proteins (i. e: the tertiary structure), having a hydrophilic surface that keeps them water soluble but a hydrophobic interior. The attractive interactions between the Paclitaxel molecule and the surrounding PEO chain could even explain the reduced hydrodynamic radius observed in PP7 (5000 g/mol) compared with PEOS (5000 g/mol). This interpretation has recently been supported by investigations of the surface tension and the viscosity of aqueous solutions of PP7[17,18]. This would have consequences for the targeting effect of the polymer mentioned above and probably also on the reaction rate of the esterase. Therefore, a broader concentration range in pure aqueous solution has to be investigated and different methods of binding Paclitaxel to the polymer have to be examined in the future in order to understand shape, transport and reaction mechanism of Paclitaxel in the complex environment where it is finally supposed to be bioavailable and active.

Acknowledgements

The Korean Science Foundation (KOSEF), Seoul, and the Deutsche Forschungsgemeinschaft (DFG), Bonn, are greatly acknowledged for their financial support, which enabled the cooperation of the laboratories at the Chosun University and the University of Duisburg. The author (M.H.) is indebted to Dr. Dirk Schubert, Freudenberg Forschungsdienste KG, Weinheim, Germany, and Dr. Jan. K. G. Dhont and Dr. Remco Tuinier, Forschungszentrum Jülich, IFF-Soft Matter, 52425 Jülich, Germany, for worthwhile discussions, C. P. Lee, GwangJu, Korea, for encouraging discussions, and the referees for their helpful comments.

[1]   N. P. Desai, P. Soon-Shiong, P. A. Sandford, *U. S. Pat.* 2,648,506 (**1997**)
[2]   B.-W. Jo, *Korean Pat.* 2000-0019873 (**2000**)
[3]   M. Hess, M. Zähres, Byung-Wook Jo, *Mater. Res. Innov.* (**2003**) in press
[4]   P. T. Callaghan, M. E. Komlosh, M. Nyden *J. Magn. Reson.* **1998**, *133*, 177
[5]   E. D. Steijskal, J. E. Tanner, *J Chem Phys* **1965**, *42*, 288
[6]   G. Otting, *Progr. Nucl. Magn. Reson. Spectr.* **1997**, *31*, 259
[7]   P. Stilbs, *Progr. NMR Spectr.* **1987**, *19*, 1
[8]   R. Kimmich, *NMR Tomography, Diffusometry, Relaxometry*, Springer Berlin, **1997**
[9]   Dhont, J K G, *An Introduction to Dynamics of Colloids*, Elsevier, Amsterdam, **1996**
[10] Russel, W B, Saville D A, Schowalter W R, *Colloidal Dispersions*, Cambridge, University Press, Cambridge, **1989**
[11] C. W. Pyun, M. Fixman, *J. Chem. Phys.* **1964**, *41*, 937
[12] S. Hanna, W. Hess, R. Klein, *Physica* **1982**, *111 A*, 181
[13] J. K. G. Dhont, M. P. B. van Bruggen, W. J. Briels, *Macromolecules,* **1999,** *32*, 3809
[14] Jo, B.-W. and Kolon Inc., PCT/Kr01/00168
[15] A. Faraone, S. Magazù, G. Maisano, P. Migliardo, E. Tettamanti, V. Villari, *J. Chem. Phys.* **1999**, *110*, 1801
[16] M. Venkatesan, C. S. Hirzel, R. Rajagopalan, *J. Chem. Phys.,* **1985**, *82*, 5685
[17] J.-S. Sohn, S.-K. Choi, B.-W. Jo, M. Hess, *Macromol. Symp.* **2003** accepted
[18] J.-S. Sohn, S.-K. Choi, B.-W. Jo, M. Hess, *Mater. Res. Innov.* submitted

The Use of Pressure-Volume-Temperature Measurements in Polymer Science

Dedicated to Prof. *Jung-Il Jin*, Korea University, Seoul, on the occasion of his 60th birthday

Michael Hess

Department of Physical Chemistry, Gerhard-Mercator-University, 47048 Duisburg, Germany

Summary: The importance of the knowledge of thermodynamic parameters from p-V-T experiments is shown in different fields of polymer science with special emphasis on determination of the phase behaviour of polymer systems.

Keywords: phase behavior; thermodynamics

Introduction

Physical chemistry describes the state of a system regardless of its shape size and constituents by variables of state like pressure p, thermodynamic temperature T, volume V and composition. The composition of a condensed material is usually given, e.g., as molar concentration c_i [mol·L^{-1}], molal concentration m_i [mol·kg^{-1}], molar fraction x_i, mass fraction χ_i, or volume fraction φ_i. The volume is sometimes normalized to the molar volume $\overline{V}$ [cm^3·mol^{-1}] or the specific volume V_{spec} [cm^3·g^{-1}], which is identical with the reciprocal density ρ^{-1} [m^3·kg^{-1}]. The variables of state are correlated by appropriate equations of state (EOS). A well-known example is the equation of state of an ideal gas

$$p \cdot V = n \cdot R \cdot T \tag{1}$$

$$R = N_L \cdot k_b = 8,31451 \; J \cdot mol^{-1} \cdot K^{-1} \tag{2}$$

DOI: 10.1002/masy.200451027

p ≡ pressure [Pa], V ≡ volume [m³], n_i ≡ amount of substance component i [mol], R ≡ gas constant, T ≡ thermodynamic temperature [K], N_A ≡ Avogadro's constant (6.022·10²³ mol⁻¹), k_b ≡ Boltzmann's constant (1.38066·10²³ J·K⁻¹).

In liquid and solid systems it is convenient to use reduced variables (indicated by ~) and reducing, characteristic values (indicated by *). The characteristic terms T*, p* and V* are correlated with molecular structures, as shown below, Equation (11) - (14).

There are extensive variables (respectively extensive equations of state) – like volume or mass (entropy S or internal energy U) – which depend on the amount of material and intensive variables (respectively intensive equations of state) – which do not depend on the amount of material – like pressure, temperature or concentration (molar entropy or density). A simple fictive experiment discriminates between the two types of physical entities: divide the system, if the variable or equation behaves correspondingly, the entity was an extensive one, if it remains unchanged it is an intensive entity.

In case of a pure substance the variables p, V and T are frequently used while in multicomponent systems the composition has to be taken into consideration, so that the state Z of a particular system consisting of i independent components can be described by a general equation of state Z(p, υ, T, n_i) with the corresponding total differential

$$dZ = \left(\frac{\partial Z}{\partial V}\right)_{T, n_i} dV + \left(\frac{\partial Z}{\partial T}\right)_{V, n_i} dT + \sum_i \left(\frac{\partial Z}{\partial n_i}\right)_{V, T, n_{j \neq i}} dn_i \qquad (3a)$$

$$dZ = \left(\frac{\partial Z}{\partial p}\right)_{T, n_i} dp + \left(\frac{\partial Z}{\partial T}\right)_{p, n_i} dT + \sum_i \left(\frac{\partial Z}{\partial n_i}\right)_{T, p, n_{j \neq i}} dn_i \qquad (3b)$$

Usually, some of the variables are held constant during experiments so that Equation 3 becomes more simple. Many physico-chemical experiments are carried out isothermally, isobarically or at constant volume without chemical reactions. Many chemical reactions are carried out at constant temperature (i e in a boiling solvent) and at constant pressure (ambient pressure).

The partial derivatives of the different fundamental equations of state are frequently easy to obtain experimentally or can be calculated by the Maxwell-relations. For the condensed materials the partial derivatives:

heat capacity at constant pressure $\qquad c_p = \left(\dfrac{\partial H}{\partial T}\right)_{p,n_i}$ (4)

heat capacity at constant volume $\qquad c_v = \left(\dfrac{\partial U}{\partial T}\right)_{V,n_i}$ (5)

coefficient of isobaric expansivity $\qquad \alpha = \dfrac{1}{V_o}\left(\dfrac{\partial V}{\partial T}\right)_{p,n_i}$ (6)

coefficient of isothermal compressibility $\qquad \kappa = -\dfrac{1}{V_o}\left(\dfrac{\partial V}{\partial p}\right)_{T,n_i}$ (7)

are of particular interest concerning processability, miscibility, phase transitions, and surface properties of polymeric materials. V_o is the initial volume of the sample, the individual partial differential quotients are called expansivity and compressibility, respectively. Heat capacities are easily accessible by calorimetric measurement, α and κ can be obtained by p-V-T-experiments. Application of Euler's chain law finally yields:

$$\left(\frac{\partial p}{\partial T}\right)_{V,n_i} = -\frac{\left(\dfrac{\partial V}{\partial T}\right)_{p,n_i}}{\left(\dfrac{\partial V}{\partial p}\right)_{T,n_i}}$$ (8)

Thermodynamic data like these and the knowledge of an appropriate equation of state are extremely useful and frequently underestimated.

A typical p-V-T experiment showing different possible transitions – including mesophase transitions is shown in Figure 1. In liquids it is frequently observed that the function V(T) is approximately linear over a rather broad range which means that α is constant. This linearity is, however, only observed when the system rests in internal equilibrium. If a glassy transition occurs or dissolved gas is liberated a sharp deviation from the linearity (a bend in the curve) is observed.

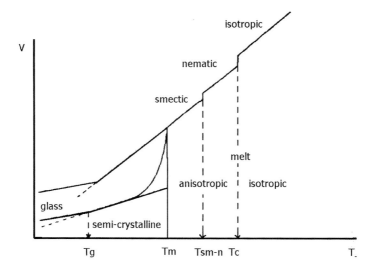

Figure 1. Schematic p-V-T diagram of an amorphous and a semi-crystalline substance, respectively, showing a glass transition and several first order (mesophase) transitions. Beyond the melting transition Tm the system obtains an (anisotropic) liquid crystalline state and the isotropic molten state at the clearing temperature Tc

Experimental Technique

There are two principally different methods of performing p-V-T measurements. The first method encloses the sample in a cylinder where the pressure is applied by a piston[1, 2, 3]. Since any leakage has to be avoided, it is necessary that cylinder and piston must fit as good as 10^{-4} cm over the whole pressure and temperature range. This method only provides a hydrostatic-type pressure equal from all sides to liquid samples and gases. The volume changes are calculated from the piston displacement. The second method has the sample in a pressure cell confined by a fluid – usually mercury. This provides an allround hydrostatic pressure to all types of samples, even for glassy and crystalline samples. Changes of the specific volume are determined from the combined changes of sample and confining fluid, usually monitored by the longitudinal displacement of a metal bellows. A commercial available equipment developed and produced by Gnomix® (P. Zoller) Boulder, Colorado, U.S.A., basing on the second principle is shown in Figure 2 and extensively described by Zoller.[4]

The external pressure is applied to a rigid sample cell, containing sample and confining fluid, which is connected to a flexible bellows. The bellows adjusts its volume according to the pressure applied by an external pressure line operated with a high-temperature stable silicone oil. The sample cell itself is surrounded by a solid, temperature adjustable pressure cell. The measuring range is from ambient temperatures to about 350°C, only limited by the boiling point of the confining fluid, and from 10 MPa to 200 MPa with an accuracy better than $2 \cdot 10^{-3}$ cm^3 g^{-1}. Zero-pressure values are obtained by extrapolation. The usual mode is isothermal, that means that the sample is set to a certain temperature and the pressure is increased in well defined, programmed steps to the end-temperature from where it goes back again to the starting temperature. The next temperature is then obtained and the pressure cycle is started again. A typical pressure step width is, e g, 10 MPa. The measuring frequency can be so small that the values often come close to equilibrium values. Typical rates for a temperature change are about 15 K/20 min and from ambient pressure to 200 MPa with a reading every 10 Mpa. A typical rate for the pressure change is 200 Mpa/5 min. The linear volume change of the bellows is measured through the displacement of a linear variable differential transformer (LVDT). The off-zero values of the LVDT are directly used for determination of the volume change with an accuracy of 10^{-3}cm^3. Isobaric experiments are also possible. The isotherms with p equal to zero are obtained by extrapolation. Near transitions it can take some more time to obtain a constant reading.

366

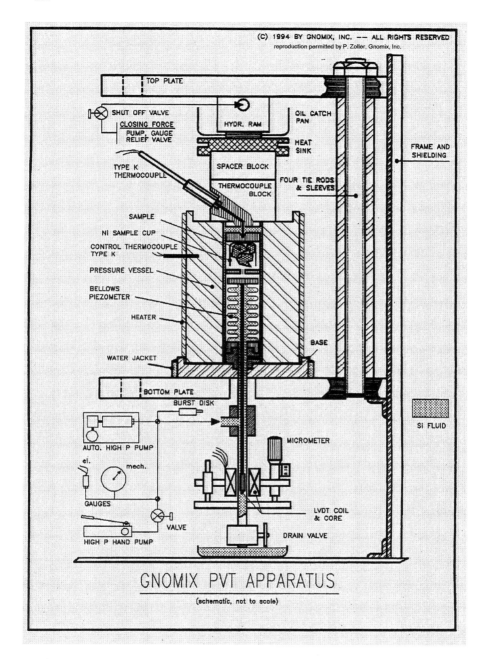

GNOMIX PVT APPARATUS

(schematic, not to scale)

Figure 2. Schematic representation of a Gnomix® machine, Gnomix Research, Boulder Coplorado, U. S. A., for p-V-T measurements. Reproduced with the permission of P. Zoller, Boulder Colorado. For details see text and the corresponding publications by P. Zoller, e.g. [4] or others by this author

Applications

p-V-T relationships in polymer solids and melts are still underestimated in their value in materials science. The data can be extremely useful in at least five major areas, such as:

- Optimizing processing parameters[a] instead of establishing such parameters by trial and error;
- Prediction of polymer-polymer miscibility;
- Correlation of the reducing parameters of equations of state (EOS) with molecular structures;
- Prediction of service performance and service life of polymeric materials and components on the basis of free volume concepts;
- Calculating of surface tension of polymer melts;
- Evaluation of start and progress of chemical reactions in polymer melts in cases when volume effects accompany the reaction;
- Materials properties of systems in contact with solvents or gases;
- Investigation of the nature of phase transitions

In the injection moulding process time and path dependence of the specific volume, phase transitions and miscibiltiy behaviour of blends play a major role during the solidification process of the molded part. If this behaviour can be simulated in a small-scale equipment, much time, material and money can be saved.[a] Menges[5, 6] and his group have designed a corresponding equipment. If material is processed which can show (different types of) mesophases, see Figure 1, the knowledge of the conditions for their existence is crucial. In some cases, however, the slow cooling rate given by the high mass of the pressure vessel, see Figure 2, may be of some disadvantage.

Polymer-polymer miscibility is governed by the Gibbs energy of mixing. It can be related to the differences between the expansivities of the components and can be described largely on the basis of free volume statistical mechanical theory of Flory and others.[7, 8, 9, 10] A low free volume means a low expansivity and a high value of the reducing temperature T^*, see below. Brostow and coworkers,[11] for example, found that V^* goes symbatically with the length of

---

[a] However, the time scale of the experimental conditions frequently differ significantly from the conditions in e g an injection moulding machine. Nevertheless many important and valuable information can be drawn from conventional p-V-T-experiments.

soft segments in polyurethanes while T* decreases with the length of the soft segments because of the corresponding weakening of interchenar interactions. For further details see also the review of Olabisi[12] and Sanchez.[13] The chemical potentials μ_i of the individual components can be derived from the reducing parameters* of an EOS, and the polymer-polymer interaction parameter χ_{23} is given in the simplest possible case by a rather crude approximation for non-polar systems:

$$\chi_{23} = p_1{}^* + p_2{}^* - 2(p_1{}^* \cdot p_2{}^*)^{1/2} \tag{9}$$

Only small differences – such as ~8 J·cm^{-3} in p* and ~190K in T* - can determine miscibility or immiscibility. However, the reduced pressure seems to be the more important value.[14] Figure 3 shows the isobars near the melting transition of a low density polyethylene, LDPE. Figure 4 shows the glass transition of a miscible blend.

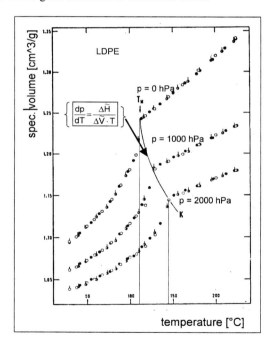

Figure 3. The melting transition of a LDPE at three different pressures after P. Zoller[15]

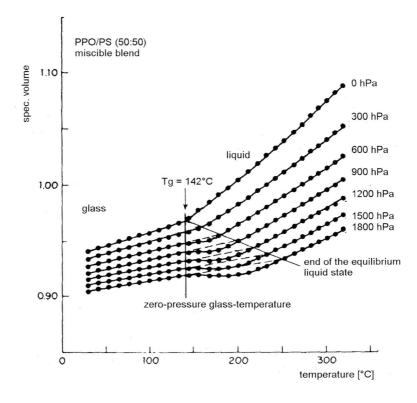

Figure 4. Pressure-dependence of the (single) glass transition of a miscible blend PPO/PS (50:50), after Zoller and Hoehn[16]

The correlation of phenomena in a p-V-T (isobaric) diagram is shown in Figure 5 with an example of a polymer liquid crystal, which has been investigated by Frenzel.[17]

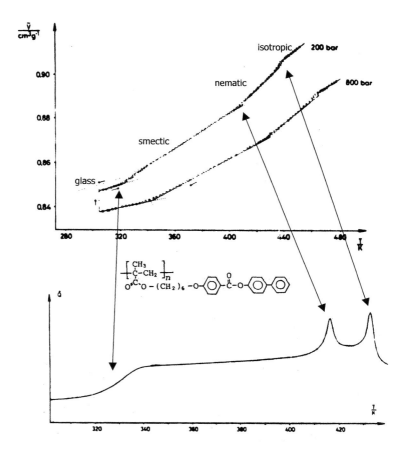

Figure 5. Phase transitions in a polymer liquid crystal p-V-T-measurement (top) compared with differential scanning calorimetry (DSC) (bottom) after Frenzel [17]

The free volume concept is very important for the interpretation of many polymer properties, see, e g, Ferry,[18] Simha,[19] Flory[20, 21] and others.[22, 23, 24, 25] Simha once stated: "if there was no 'free volume' it has to be invented".[26] The free volume V^f is defined by:

$$V \equiv V^* + V^f \tag{10}$$

with $V^f \equiv$ free volume and $V^* \equiv$ the characteristic volume for a given equation of state. Positron annihilation measurements can give an estimate of V^f. V^* is also sometimes called the 'net volume'[27] or the hard-core volume[28, 8, 29]. Hence, V^* can be interpreted as the volume that is left after the free volume has been squeezed out of a system. The free volume depends on temperature, pressure, stress level in tension, compression, shear, bending etc, and frequency in oscillatory experiments. The term:

$$\tilde{V} \equiv \frac{V}{V*} \tag{11}$$

is called reduced volume. Reduced values, indicated by the tilde (~) and characteristic values, indicated by *, occur in equations of state (EOS). Important values are the reduced pressure $\tilde{p}$ and the reduced temperature $\tilde{T}$, normalized correspondingly to Equation (11). The physical meaning of the characteristic (reducing) values p^* and T^* are not as "simple" as the interpretation of V^* and may differ in the different theories. In terms of Hartmann's approach with

$$\tilde{p} \cdot \tilde{V}^5 = \tilde{T}^{1.5} - \ln \tilde{V} \tag{12}$$

p^* is given by:

$$p^* = \frac{2 \pi N}{9 \mu V} r \frac{d(u)r}{dr} g(r) \tag{13}$$

$\frac{N}{V}$ is the number of segments per unit volume, μ is a geometrical factor depending on the coordination number z ($z = 6 \rightarrow \mu = 1$), $u(r)$ is the interaction potential, and $g(r)$ is the binary radial distribution function with the average segment distance r.

$$T^* = \frac{3(z-2)\varepsilon^*}{k} \tag{14}$$

ε^* is the intersegmental potential energy minimum, hence T* is high when the intersegmental interactions are strong. k is the Boltzmann constant. Figure 6 shows an example of the description of a series of polymers with the Hartmann equation.

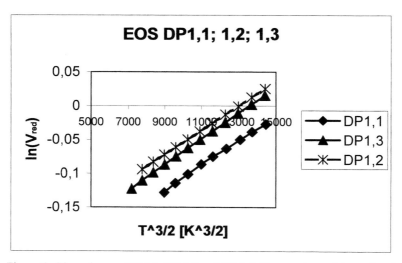

Figure 6. The polymers DP 1,1, DP 1,2 and DP 1,3 (the structure of the molecules is given in Figure 7, the number gives the number of carbons in R1 and R2, respectively) can be described in the fluid state by the Hartmann equation, Equation (12). The "wiggles" in the graph of DP 1,1 reflect mesophase transitions which are different and much weaker in DP 1,2 and DP 1,3, respectively.

The reduced volume, for example, is used in the description of the temperature shift factor a_T of a pre-drawn sample, which is given by Brostow et al. [11]:

$$\ln a_T = 1/(a + c\lambda) + B/(\tilde{V} - 1) \tag{15}$$

a, B and c are parameters, λ is the draw ratio. The reduced volume $\tilde{V}$ depends on T* via an EOS, while T* in turn depends on λ. Brostow[11] and also Hartmann[27] could show that the shift factor a_T obtained from p-V-T measurements is in good agreement with the results from other experimental methods. The reducing values can be determined from p-V-T measurements using an appropriate EOS, as for example the Hartmann equation,[30] Equation (12), which reduces to the Simha-Somcynsky equation at zero pressure. The definition of the reducing parameters may be different according to the EOS used. V, α, and κ are required to describe liquids in terms of Flory's theory of the liquid state.[7-9] Another still widely used approach to p-V-T equations of state is the relationship by van-der-Waals, and it has been shown by Roszkowski[31] that it can be obtained by simplification of Flory's equation of state. Other EOS were provided for example by Sanchez and Lacombe,[32] and Curro.[33] Simha's hole theory[34, 35, 36] extended to melts of compatible blends,[37] are able to describe experimental data over large spans of temperature and pressure. There are also some quite successful semi-empirical EOS, such as the Tait equation[38] for polymer melts, however, the Tait equation is not valid for semi-crystalline polymers.

$$V(p, T)/V(p = 0, T) = 1 - 0.0894 \ln [1 + p/B(T)] \qquad (16)$$

with
$$B(T) = B_0 \exp(-B_1 T) \qquad (17)$$

Bends in $V(T)_p$ graphs indicate phase transitions and the onset of molecular motions, the shift factor of relaxation processes and the activation energy can be calculated and give information about important structure-property relations.[39]

Surface tension γ of polymer melts is sometimes diff, there are relationships between other thermodynamic parameters and the surface tension or the interfacial free energy γ_{ii} between two condensed phases. Amóros et al. [40, 41] have studied the relationship between $\kappa(T)$ and the surface free energy. There are also theoretical approaches enabling the computation of γ from p-V-T data. These are based on Van-der-Waals' principle of corresponding states, which was extended by Prigogine et al. [42, 43, 44] to chain molecules. Patterson and Rastogi,[45] Siow and Patterson[46] and Dee and Sauer[47] developed equations to determine γ from p-V-T icult to determine experimentally. Some of the techniques are time consuming or require fairly elaborated equipment, e g, the 'imbedded fibre retraction method'.[48] However data respectively from the reducing *-parameters mentioned above using one of the appropriate EOS. Berry, Brostow and Hess[49] have successfully applied the theories on longitudinal polymer liquid crystals with varying mesogen concentration.

Breuer and Kosfeld[50] investigated the p-V-T properties of mixtures of polymers with solvents and gases. The behaviour of these systems is extremely important for their use as sealant and gasket or in the production of foams. They found that a linear relation between the compressibility and the pressure and the temperature, respectively, is only given when the system is in equilibrium. Sharp deviations were observed at the glass transition and when dissolved gas was liberated.

Starting from accurate compressibility measurements on atactic PS in the freezing region it was found by Rehage and coworkers[51, 52, 53, 54] that the relations for a second order transition are not applicable since the thermodynamic properties in the glassy state not only depend on temperature and pressure but also on the way of the glass formation, see also Hartmann and Haque.[27] This also applies to the limiting cases of very high pressures and temperatures (high pressure does NOT lead to a second order transition) and infinitely slow cooling at normal pressure (Gibbs-DiMarzio theory[55, 56] predicts a second order transition in this case). The path-dependence can be described by internal order parameters. Phase- and kinetic measurements have led to different conclusions about the multiplicity of linear independent order parameters. Later the same was found for polymer liquid crystals with the mesogenic groups on the side-chain [17] and as a constituent of the polymer backbone.[57, 58] Increasing pressure enlarges the range of existence of mesophases. Incorporation of rigid or mesogenic groups in the polymer chain does not change the character of the glass transition. The Ehrenfest[59] equations should not be valid at the glass transition that cannot be described as a second order transition.

While first order transitions obey the Clausius-Clapeyron equation:

$$\left(\frac{dp}{dT}\right)_{trans} = \frac{\Delta \tilde{S}}{\Delta T} = \frac{\Delta \tilde{H}}{\Delta \tilde{V} \cdot T} \tag{18}$$

The Ehrenfest equations are:

$$\left(\frac{dp}{dT}\right)_{trans} = \frac{\Delta c_p^*}{T \cdot \Delta \alpha^*} \tag{19a}$$

$$\left(\frac{dp}{dT}\right)_{trans} = \frac{\Delta \alpha^*}{\Delta \kappa^*} \tag{19b}$$

combining Equation (19a) and (19b) yields:

$$\frac{\Delta \alpha^*}{\Delta \kappa^*} = \frac{\Delta c_p^*}{T \cdot \Delta \alpha^*} \tag{20}$$

Rehage and Borchard[50] have shown that it is necessary to introduce parameters ξ describing the degree of internal order in a glass. If there is only one order parameter ξ necessary, either Equation (19a) **and** (19b) are valid or none of them. If only one of the Equations 19 is found to be valid at a glass transition this means that more than one parameter ξ has to be introduced. Meixner[60] and Davies and Jones[61] had already earlier predicted that:

$$\frac{\Delta c_p \cdot \Delta \kappa\dot{}}{T \cdot \left(\Delta \alpha^*\right)^2} \geq 1 \qquad (21)$$

This was also verified for the polymers DP 1,1 to DP 1,3.[62] The pressure dependence of the glass transition temperature is usually found in the range from $dT/dp = 0.2$ K/MPa to $dT/dp = 0.8$ K/MPa. The pressure dependence of the glass transition temperature is not necessarily linear, see Figure 3. Increasing solidification pressure frequently leads to a densification of about 1% per 100 MPa. The pressure dependence of first order transitions are generally found to be around 0.2 K/MPa to 1 K/MPa.

Anomalous behaviour of melts can be analysed such as pressure induced melt-crystallization[63] or monotropic mesophase transitions.[54-56] Figure 7 and Figure 8 show an example. p-V-T measurements can also gain insight into the nature of certain phase transitions such as disordering processes hidden by melting processes as has been shown by Cantow and coworkers[64].

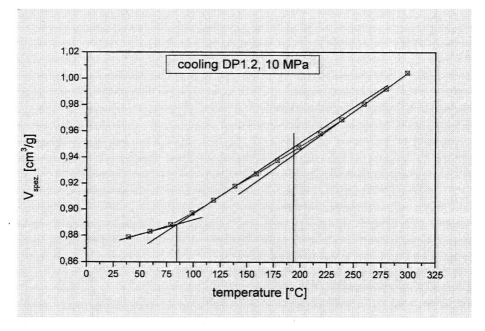

poly[oxy(2,2'-dialkylpropane-1,3-diyl)carboxybisphenyl-4,4'-dicarbonyl]

Figure 7. An example for an unusual phase transition: DP 1,2 shows an unusual increase of the specific volume with decreasing temperature, just as water at its melting point. The numbers indicate the number of carbons in the side chains

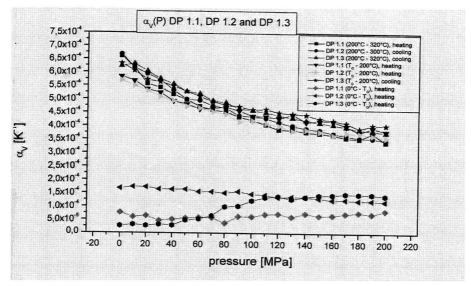

Figure 8. Unusual transition in the glassy state of the polymer DP 1,3 (-•-, bottom) compared with DP 1,1 and DP 1,3. The transition occurs in the pressure-dependence of the expansivity and is probably caused by packing effects of the growing side chain. The group of curves at the top describes the melt, the group of curves at the bottom describes a glass. In the case of DP 1,3 an increasing α with the pressure is observed. Rastogi[65] has recently described similar effects in poly(4-methyl-pentene-1).

Acknowledgement

The author expresses his thanks to The Deutsche Forschungsgemeinschaft, DFG (Bonn, Germany) and KOSEF (Seoul, South Korea) for supporting the sabbatical at Chosun University, Kwang-Ju, South Korea, (hosted by Prof. Byung-Wook Jo) during which a great part of this manuscript was written. Dr. Jürgen Pionteck and his team at the Institute for Polymer Research e. V., Dresden, Germany, were very helpful with some experiments, finally, C. P. Lee (Cambridge, U. K.) is thanked for helpful discussions.

378

[1]   H. Breuer, G. Rehage, *Kolloid Z* **1967**, *159*, 216.
[2]   G. Menges, P. Thienel, *Kunststoffe* **1975**, *65*, 698.
[3]   G. N. Foster, R. G. Griskey, *Rev. Sci. Instrum.* **1964**, *41*, 759.
[4]   P. Zoller, P. Bolli, V. Pahud, H. Ackermann, *Rev. Sci. Instrum.* **1976**, *47*, 948.
[5]   S. Stitz, PhD Thesis, RWTH Aachen, Germany, 1973.
[6]   J. Vargel, PhD Thesis, RWTH Aachen, Germany, 1974.
[7]   P. J. Flory, R. A. Orwoll, A. Vrij, *J. Am. Chem. Soc.* **1964**, *86*, 3507.
[8]   P. J. Flory, *J. Am. Chem. Soc.* **1965**, *87*, 1833
[9]   P. J. Flory, *Disc. Faraday Soc.* **1970**, *49*, 7.
[10]   D. J. Walsh, S. Rostami, *Adv. Polym. Sci.* **1985**, *70*, 119.
[11]   W. Brostow, J. V. Duffy, G. F. Lee, K. Madejczyk, *Macromolecules* **1991**, *24*, 479.
[12]   O. Olabisi, L. M. Robeson, M. T. Shaw, *"Polymer-Polymer Miscibility"*, Academic Press, New York 1974.
[13]   I. Sanchez, in: *"Polymer Blends"*, Chapter 3, D. Paul, S. Newman, Eds., Academic Press, NewYork 1978.
[14]   D. J. Walsh, W. W. Graessley, S. Datta, D. J. Lohse, L. J. Fetters, *Macromolecules* **1992**, *25*, 5236.
[15]   P. Zoller, *J. Appl. Polym. Sci.* **1979**, *23*, 1051.
[16]   P. Zoller, H. H. Hoehn, *J. Polym. Sci.: Polym. Phys.* **1982**, *20*, 1385.
[17]   J. Frenzel, PhD Thesis, Clausthal, Germany, 1981
[18]   J. D. Ferry, *"Viscoelastic Properties of Polymers"*, 3rd edition, Wiley, New York 1980.
[19]   R. Simha, T. Somcynsky, *Macromolecules* **1969**, *2*, 342.
[20]   P. J. Flory, *J. Am. Chem. Soc.*, **1965**, *87*, 1833.
[21]   P. J. Flory, *Disc. Faraday Soc.* **1970**, *49*, 7.
[22]   W. Holzmüller, *Adv. Polym. Sci.* **1978**, *26*, 1.
[23]   S. Matsuokain, in: *"Failure in Plastics"*, Chapter 3, W. Brostow, R. D. Corneliussen, Eds., Hanser, Munich 1986.
[24]   J. Kubat, M. Rigdahl, in: *"Failure in Plastics"*, Chapter 4, W. Brostow, R. D. Corneliussen, Eds., Hanser, Munich 1986.
[25]   L. C. E. Struik, in: *"Failure in Plastics"*, Chapter 11, W. Brostow, R. D. Corneliussen, Eds., Hanser, Munich 1986.
[26]   R. Simha, private communication.
[27]   P. J. Flory, R. A. Orwoll, A. Vrij, *J. Am. Chem. Soc.* **1964**, *86*, 3507.
[28]   B. E. Eichinger, P. J. Flory, *Trans. Faraday Soc.* **1968**, *64*, 2035.
[29]   W. Brostow, *Macromolecules* **1971**, *4*, 742.
[30]   B. Hartmann, M. Haque, *J. Appl. Phys.* **1985**, *58*, 2831.
[31]   Z. Roszkowski, *Mater. Chem. Phys.* **1981**, *6*, 455.
[32]   I. C. Sanchez, H. Lacombe, *J. Polym. Sci. Polym. Phys. Ed.* **1977**, *15*, 71.
[33]   J. G. Curro, *J. Macromol. Sci. Rev. Macromol. Chem.* **1974**, *11*, 321.
[34]   R. Simha, *Macromolecules* **1977**, *10*, 1025.
[35]   R. K. Jain, R. Simha *Macromolecules* **1980**, *13*, 1501.
[36]   R. Simha, R. K. Jain, *J. Polym. Sci. Polym. Phys. Ed.* **1978**, *16*, 1471.
[37]   R. K. Jain, R. Simha, P. Zoller, *J. Polym. Sci. Polym. Phys. Ed.* **1982**, *20*, 1399.
[38]   G. P. Tait, *"Physics and Chemistry of the Voyage of H. M. S. Challenger"*, Vol. 11, Pt. IV, Sci. Papers LXI (1888); Vol. II,1, University Press, Cambridge 1900.
[39]   J. Sivergin, S. M. Usmanov, V. V. Amirov, S. M. Kireeva, *Plaste & Kautschuk* **1994**, *41*, 168.
[40]   J. Kirjava, T. Rundkvist, R. Holsti-Mietinnen, M. Heino, T. Vainio, *J. Appl. Polym. Sci.* **1995**, *55*, 1069.
[41]   J. Amóros, J. R. Solana, E. Villar, *Mater. Chem. Phys.* **1982**, *7*, 127.
[42]   J. Amóros, J. R. Solana, E. Villar, *Mater. Chem. Phys.* **1982**, *7*,767.
[43]   I. Prigogine, *J. Chim. Phys.* **1950**, *47*, 807.
[44]   I. Prigogine, L. Sarolea, *J. Chim. Phys.* **47**, 807 (1950)
[45]   I. Prigogine, A. Bellermans, V. Mathot, *"The Molecular Theory of Solutions"*, North-Holland, Amsterdam 1957.
[46]   D. Patterson, A. K. Rastogi, *J. Phys. Chem.* **1970**, *74*, 1067.
[47]   K. S. Siow, D. Patterson, *Macromolecules* **1971**, *4*, 27.
[48]   G. T. Dee, B. B. Sauer, *Polymer* **1995**, *36*, 1673.
[49]   J. M. Berry, W. Brostow, M. Hess, *Polymer* **1998**, *39*, 4081.
[50]   H. Breuer, R. Kosfeld, *Adv. in Chem. Ser.* **1965**, *48*, 125.
[51]   H. Breuer, G. Rehage, *Kolloid Z. Z. Polymere* **1967**, *216*, 159.
[52]   H.-J. Oels, G. Rehage, in: *"The Physics of Non-Crystalline Solids"*, G. H. Frischat, Ed., Aedermannsdorf 1977, p. 411.
[53]   H.-J. Oels, G. Rehage , *Macromolecules* **1977**, *10*,1036.
[54]   G. Rehage, W. Borchard, in: *"The Physics of the Glassy State"*, R. N. Haward, Ed., London 1973, p. 54.

[55]   H. Gibbs, E. A. DiMarzio, *J. Chem. Phys.* **1958**, *28*, 373.

[56]   E. A. DiMarzi, H. Gibbs, P. D. Fleming, *Macromolecules* **1976**, *9*, 763.

[57]   M. Hess, J. Pionteck, *Mater. Res. Innov.* **2002**, *6(1)*, 31.

[58]   M. Hess, *Mater. Res. Innov.* **2002**, *6(1)*, 31.

[59]   P. Ehrenfest, *Proc. Kon. Akad. Wetensch. Amsterdam* **1933**, *36*, 153.

[60]   J. Meixner, *Compt. Rend.* **1952**, 432.

[61]   R. O. Davies, G. O. Jones, *Adv. Phys.* **1953**, *2*, 370.

[62]   R. Woelke, PhD Thesis Duisburg, Germany, 2001.

[63]   D. Walsh, T. Ougizawa, W. H. Tuminello, K. H. Gardener, *Polymer* **1992**, *33*, 4793.

[64]   H. Lill, B. Rudolf, H.-J. Cantow, *Polymer Bulletin* **1993**, *30*, 231.

[65]   S. Rastogi, G. W. H. Höhne, A. Keller, *Macromolecules* **1999**, *32*, 8897.

RETURN TO: **CHEMISTRY LIBRARY**

100 Hildebrand Hall · 510-642-3753

LOAN PERIOD	1	2		3
4		5		6

2-HR USE

ALL BOOKS MAY BE RECALLED AFTER 7 DAYS.

Renewals may be requested by phone or, using GLADIS,
type **inv** followed by your patron ID number.

DUE AS STAMPED BELOW.

FORM NO. DD 10
3M 5-04

UNIVERSITY OF CALIFORNIA, BERKELEY
Berkeley, California 94720–6000